Tropical Fish as a Hobby

by Mary E. Sweeney

SAVE-OUR-PLANET SERIES

T.F.H. Publications, Inc.
1 T.F.H. Plaza • Third & Union Aves. • Neptune, NJ 07753

Contents

Introduction 2
Equipment 5
Water of Life 21
Aquarium Set-Up 25
Foods and Feeding 37
Maintenance 44
The Fishes 53
Fish Health 89
Suggested Reading 95
Index 95

DEDICATION

For Eamonn, the little boy
for whom I would like the planet saved.

Photography & Illustrations: Dr. Herbert R. Axelrod, M. Batell, K.L. Chew, B. Degen, J. Elias, M. Gilroy, Dr. H. Grier, M. Jacobs, B. Kahl, H. Linke, A. van den Nieuwenhuizen, L. O'Connell, W. Petersmann, M.P. & C. Piednoir, J. Quinn, H.-J. Richter, M. Roberts, W. Ross, A. Roth, V. Serbin, M. Smith, W. Sommer, A. Spreinat, R. Stawikowski, K. Tanaka, W. Tomey, J. Vierke, L. Wischnath.

A crisply marked *Metynnis* sp., one of the fishes commonly known as "silver dollars." Photo by Aaron Norman.

Introduction

One of the most popular hobbies in the world today is the keeping of tropical fish as pets. They don't bark, shed, need to be taken for walks, or cause trouble with the landlord. They don't cost a lot, send allergy sufferers into a sneezing frenzy, or talk back. Tropical fish can give you hours of pure, simple pleasure. The quiet, watery world of fishes is one of activity and charm. There is always something going on. The guppy is waiting to deliver some babies, the catfish is looking for food in the gravel, and the betta is busy looking out at you. The tetras are swimming together in a school and the danio and the swordtail are playing tag. And you can have so many more fish than you could ever have of any other kind of pet—except maybe if you had an ant farm.

A tropical fish aquarium is also tremendously helpful in managing one of the most common and life-threatening conditions of modern life...stress. Researchers are now telling us that *watching* fishes also contributes to wellness in that it is a relaxing activity that reduces blood pressure, lowers heart rate, and reduces pain and anxiety.

Have you ever wondered why so many doctors and dentists have gone to the trouble of setting up beautiful aquariums in their waiting rooms? Surely they are not merely decorative. A decoration for a busy medical or dental practice would be

987654321   **1995 Edition**   9 6789

Distributed in the UNITED STATES to the Pet Trade by T.F.H. Publications, Inc., One T.F.H. Plaza, Neptune City, NJ 07753; distributed in the UNITED STATES to the Bookstore and Library Trade by National Book Network, Inc. 4720 Boston Way, Lanham MD 20706; in CANADA to the Pet Trade by H & L Pet Supplies Inc., 27 Kingston Crescent, Kitchener, Ontario N2B 2T6; Rolf C. Hagen Ltd., 3225 Sartelon Street, Montreal 382 Quebec; in CANADA to the Book Trade by Vanwell Publishing Ltd., 1 Northrup Crescent, St. Catharines, Ontario L2M 6P5 ; in ENGLAND by T.F.H. Publications, PO Box 15, Waterlooville PO7 6BQ; in AUSTRALIA AND THE SOUTH PACIFIC by T.F.H. (Australia), Pty. Ltd., Box 149, Brookvale 2100 N.S.W., Australia; in NEW ZEALAND by Brooklands Aquarium Ltd. 5 McGiven Drive, New Plymouth, RD1 New Zealand; in Japan by T.F.H. Publications, Japan—Jiro Tsuda, 10-12-3 Ohjidai, Sakura, Chiba 285, Japan; in SOUTH AFRICA by Lopis (Pty) Ltd., P.O. Box 39127, Booysens, 2016, Johannesburg, South Africa. Published by T.F.H. Publications, Inc.

MANUFACTURED IN THE UNITED STATES OF AMERICA
BY T.F.H. PUBLICATIONS, INC.

An aquarium like this one will undoubtedly be the center of a great deal of attention.

unlikely to require partial water changes or daily feedings, or trips to the pet shop for supplies or new fish. So why then do we so often see these gorgeous displays of tropical and marine fishes? Are all these doctors bringing their hobbies into the office? Not likely. Like meditation and biofeedback, the simple act of watching fishes in an aquarium can reduce blood pressure and heart rate in both hypertensive and non-hypertensive individuals. And that's why we are now seeing aquaria in medical offices, dental offices, businesses, and even a new 3,000-gallon saltwater aquarium on the stock-trading room floor.

Researcher Lynette A. Hart, who has been working with patients in hospital Coronary Care Units, has shown that an aquarium in a hospital room reduces the anxiety and enhances the self esteem of the patient. As an added benefit, she finds that friends and relatives who visit the patients are more likely to prolong their visits when there is an aquarium in the room. Third, and most likely to impress the physicians and insurance companies alike, is the suggestion that hospital rooms with aquaria are associated with more frequent discharges to home rather than transfers within the hospital or unapproved discharges A.M.A. (against medical advice).

Physical benefits aside, keeping tropical fish as a hobby comes under the category of good, clean, wholesome fun. It appeals to all ages and indeed may even provide a link between

A planted, well stocked aquarium can be a masterpiece. The hobby has advanced markedly in recent years, and such a tank is possible for every hobbyist. Photo courtesy of American Acrylic Mfg., San Diego, California.

generations. Parent, child, or grandparent, aquariums appeal to the youngest and the oldest and are a fine example of harmony...in nature, in the aquarium, and in the body and the soul.

The aquarium of today is a pleasure to own. Onerous water changes, complicated equipment, and replacing your lost fish every week are a thing of the past. The modern aquarium is very elegant and sophisticated with high-tech acrylic tanks, fool-proof filtration, furniture-quality stands...and plenty of information for the new aquarist. A 30-gallon community tank can provide hours of entertainment and a good look at the antics of your favorite fishes. A 50-gallon Amazon biotope can duplicate the mysterious, shadowy habitat of the angelfish and the discus. Twenty gallons of water and two goldfish named Henry and June will often replace Saturday morning cartoons for the little ones. There are hundreds of different combinations of plants and fishes that will coexist in peaceful harmony in the aquarium.

The aquarium hobby has finally come into its own. Once thought to be a hobby for young Huck Finns to be cast off in time and replaced with toys for older boys, we are finding ever more evidence that the sight of calmly swimming fish evokes in all of us a sense of serenity, safety, and well-being. Fish are *good* for you! Welcome.

Equipment

The variety and availability of aquarium products to help you enjoy your hobby is sometimes overwhelming to the new fishkeeper. The following information will help guide you into making the selections that will enhance your enjoyment and your success in this new hobby.

CHOOSING THE AQUARIUM

The usual advice given to tropical fish hobbyists is that the aquarium be rectangular in shape. For housing most fishes, and for keeping most fishes in a given volume of water, this advice is sound. It would be a shame, however, if you felt strictly limited or deprived by this advice when your heart is really set on having another kind of aquarium. While the fancier shapes like hexagons and cylinders require more knowledge and perhaps special equipment to keep the fishes healthy, they can be stunning when planted and stocked with the appropriate fish, like angelfish perhaps. While a tall thin tank is out of the question if you are interested in keeping any of the fast-swimming active fishes like barbs or tetras, who would probably concuss themselves on the sides, they are lovely with a few select, sedate fish.

With a basic understanding of the needs of your fishes and the services performed by your equipment, you will be free to set up the aquarium of your dreams. What goes on in the aquarium as far as the water quality is concerned is easily

The size and shape of the aquarium you select will depend largely upon the fishes you intend to keep. Beyond that, it's a matter of personal preference.

monitored and controlled by you, the aquarist. As time and experience develop your "wet thumb," you will become adept at judging when the filter needs cleaning or the water changed. Until then, try to read as much as you can about your fishes and their care.

The aquarium you choose can be either glass or acrylic. It's strictly a matter of personal preference. Glass is generally more durable as far as scratches are concerned, but on the other hand, acrylic is available in intriguing shapes and is much lighter than glass. Glass aquaria are less expensive, enabling the aquarist on a budget to purchase a larger tank at the outset, but acrylic has a more modern, decorator look. Your fish will not care whether they are in a glass aquarium or an acrylic aquarium. Most important are size, shape, and quality. Whenever you are shopping for aquarium equipment, buy the largest and the best you can afford. Never buy a repaired leaker, or a used tank, or a discount-store tank. Only buy a well-known brand name from a reputable pet shop. The dealer who sells quality fish will fully understand the necessity of a quality tank and will provide the same. When purchasing an all-glass aquarium, make sure that it is not chipped, and that the edges are nicely beveled so that they will not cut you when the aquarium is handled.

The larger the tank is, the thicker the glass must be to withstand the tremendous pressure water can

exert. Very big tanks will have supporting struts fitted on top from front to back. All quality tanks will feature supports for a glass cover—highly recommended for limiting the amount of dust that will fall onto the water's surface and to minimize water loss through evaporation. Glass covers for large tanks should be hinged for access.

HEATERS

Tropical fishes have evolved to live in the temperatures of their native waters. Man cannot overcome the conditioning of a few million years of evolution and "train" captive tropicals to live in water kept at the temperature of his own choosing. When you buy your fish, or better yet, *before* you buy your fish, educate yourself about the basic requirements of the species...and give them these conditions! There is no excuse for keeping a fish that is meant to be in water of 75°F in significantly higher or lower water temperatures. Regardless of the relative hardiness of the fish, keeping a fish outside of its required temperature range will eventually lead to the early death of a fish to which you have probably become quite attached.

It is important to keep in mind that the larger the aquarium the greater its weight will be. Whatever the aquarium is placed on must be able to support the weight.

There are many heaters on the market, but the most popular type these days is the submersible heater controlled by a thermostat. Preset submersible heaters can be calibrated to the desired temperature and should ever after maintain that temperature.

Heaters should never be placed into the gravel; this can result in uneven heating of the glass casing, which might then shatter or at least crack and allow water to enter the heater with dire results.

In large aquaria it is always better to use two heaters, one at each end of the aquarium, than one high-wattage heater. This not only ensures more even distribution of heat, but if one heater fails then at least the temperature drop will take much longer, as it is unlikely both will fail at the same time.

The heater should be placed so that it heats the lower level of the aquarium, thus creating an upward current of warm water. Heaters can also be placed into some of the external power filters, thus forming a thermofilter, but here the problem is that if the pump fails for any reason then the heating system is out as well.

THERMOMETERS

Even though the heater may be controlled by a thermostat, you should always use an internal or external thermometer, or both, in the aquarium. A thermostat may fail and this will result in the water temperature rising to the full potential of the heater. This may just be enough to cook the fish! On the

When installing an aquarium heater, it is advisable to read the instructions before plugging it in.

other hand, it might jam in the closed position so the heater does not come on. The thermometer is therefore a very valuable, yet inexpensive, accessory.

There are numerous types of thermometers available, each having their own advocates. Some are free floating, some are standing, others are fixed inside the tank with suction cups, and yet others can be clipped to the sides of the aquarium. The most recent types are liquid crystal and are attached to the outside of the tank in a convenient, unobtrusive place. They are self-adhesive and very accurate. Do not use a mercury thermometer inside an aquarium. If it breaks, the poisonous contents will kill the fishes. Alcohol is a much better fluid for inside models.

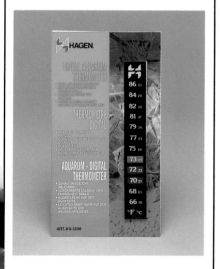

Digital thermometers can be attached to the outside of the aquarium (as in upper photo) and are easy to read. Photos courtesy of Hagen.

FILTERS

Filtration is a broad term that is used to cover all the processes that help to maintain water quality. Filtration can be divided into the following types: mechanical, chemical, biological, vegetative. Although there is usually not much deep thought devoted to the types of filtration, most aquaria enjoy to some extent the benefits of all the types of filtration. The filter you employ in your tank provides the mechanical

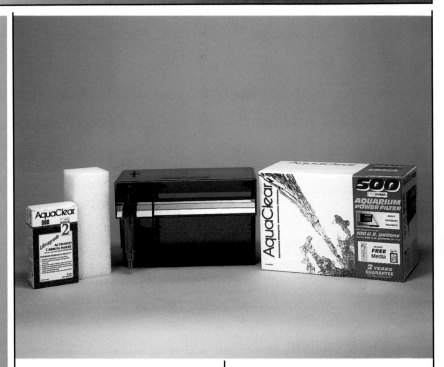

Outside power filters will help keep your water clean and your fishes healthy. Plan to purchase a filter with a flow rate sufficient for the capacity of your aquarium. Photo courtesy of Hagen.

filtration, sifting out particles of debris that will cloud your water. The carbon and other special additives, such as zeolite, provide the chemical filtration that eliminates the toxic compounds. The bacteria thriving in a mature tank provide the biological filtration which turns ammonia and nitrites into less harmful substances, and finally, the plants and algae do their jobs vegetatively on the remaining nitrates.

Unfortunately, during the initial start-up period when a lot of people get really frustrated with cloudy water conditions, the mechanical and chemical filtration actions are the only ones functioning. If you hang on, follow the few rules about aquarium start-up, and trust that you have done it properly, you will have the clear water of your dreams in just a few days.

Box Filters

The simplest filters of all are the small box filters that are made of plastic and can be situated in the corner of an aquarium. I often see them in aquaria in pet shops. A plastic tube leads from the base of the inside of the filter and an airstone is fitted into this. A suitable filter medium, such as floss, is placed into the box, which has holes or slots through which water can flow. It is also helpful to place a layer of marbles in the bottom of the box. They will prevent the filter box from floating around and provide a suitable surface area for beneficial bacteria. Other additives, such as carbon and zeolite can also be layered in the box of the filter.

Once the box is placed into the aquarium it fills with water, including the tube that opens at or just below the water surface. Once the air is switched on, it creates the upward current that draws water from the area around it, causing the debris carried into the filter to collect in the box which can be regularly removed so

that the filter medium can be replaced or cleaned. It is a mechanical filter and has moderate biological filtration properties.

Outside Power Filter

This is a very popular style of filter because of its efficiency, ease of maintenance, and reliability. It is simple to set up and simple to maintain. It creates both mechanical and biological filtration, and the addition of charcoal to the filter media provides the element of chemical filtration as well. Virtually every filter manufacturer offers his particular style, and it remains only to select the one that

Box filters are ideal for small aquariums with a capacity of 20 gallons or less.

Aquarium carbon will remove toxins from the water when you use it in your filter. Remember to change it frequently, as it does become saturated over time. Photo courtesy of Hagen.

Sponge filters provide a constant source of vital biological filtration.

suits your own taste and wallet.

Undergravel Filter

The undergravel filter uses a plate, usually plastic, that is raised about ½ inch above the bottom glass of the aquarium, coupled with airlift tubes and air supply to draw dirty water down through the filter plate. The gravel covering the filter is home for nitrifying bacteria, making this a very effective filter.

Undergravel filters are really meant for marine aquaria and I personally don't like to use them in freshwater

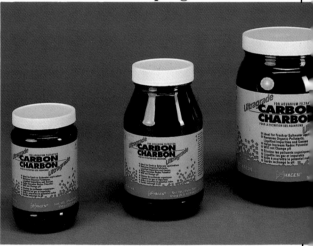

aquaria. Digging fishes and perhaps even the aquarist's personal need to clean under the filter plate can upset what is a delicate balance of bacterial action and oxygen. If you think you would like to use an undergravel filter, discuss the implications in detail with an experienced pet shop owner.

Sponge Filter

The sponge filter is often

overlooked, but is in reality extremely effective, particularly when raising fry. A sponge filter is coupled with a simple air pump. The water is drawn through the sponge and bubbles and water exit through a plastic stem. The surface of the sponge is an ideal medium for the nitrifying bacteria and will often collect particulate matter. Sponges are host to rotifers which are delicious

food for fishes. It is amazing to see just how quickly a cloudy aquarium is cleared when a mature sponge filter is introduced—and how clear it stays. The trick is to keep the bacteria going in the sponge. It's easy to do, just rinse the sponge periodically with lukewarm water, rather than hot, and make sure it's hooked up to the pump all the time.

Canister Filter

The canister filter is the workhorse of the aquarium. It seems a bit complicated in the box, but once assembled, it is really quite simple. The filter and mechanical parts are self-contained in the canister. Two hoses, one to withdraw water from the tank and one to return water to the tank, are attached to the canister. The canister is filled with layers of filter media that filter the water until it is crystal clear and biologically cleansed.

AERATION

The addition of extra air into an aquarium by mechanical means is termed *aeration*, and contributes to the well-being of your fish by increasing the circulation of the water which is necessary for the exchange of gases at the water's surface. The bubbles

produced when air is pumped into an aquarium actually contribute very little oxygen themselves. It is the fact that they carry the deoxygenated water to the surface that is instrumental in increasing the overall oxygen content in the aquarium.

If the bubble size is small, and if a curtain of bubbles is produced by the airstone, then this will add *some* oxygen, which will be released when the bubbles burst. Not only do the air bubbles take oxygen-depleted water to the surface where it can be renewed with oxygen, but they also contribute in another way. When the bubbles burst at the surface they agitate the water, thus increasing the surface area, which allows for more oxygen to dissolve into the water than would

be the case if the surface were undisturbed. With maximization of oxygen content in the tank, the fish population can also be increased. Aeration also aids in the removal of toxic gases.

The beneficial nitrifying bacteria that convert dangerous nitrites to nitrates need a good supply of oxygen, so again, aeration will have a positive effect on the aquarium's ability to remove harmful toxins.

Airstones
The normal way in which air is introduced to an aquarium is via an airstone attached by a length of plastic tubing that is connected to an air pump. An airstone can be any porous material, such as felt, stone, or wood; usually the molded stones are

used. These may be small cubes or long rectangular diffusers. The amount of air released can vary, not only by the pressure pumped in, but by the size of the airstone. It is usually placed near the back of the aquarium, often concealed behind a rock. By featuring two or more long airstones you can create a wall or curtain of air bubbles that can be very attractive.

Pumps

There is no shortage of pumps on the market and they are available in a range of outputs. These pumps are simple and relatively inexpensive, their only drawback being the level of noise they may produce. The better models are very quiet, but even noisier models can be made quieter by suspending them on a hook so that they are not in contact with any solid surface. This would also overcome the tendency of some of these pumps to "wander" across the unit on which they were placed. Diaphragm pumps are quite able to cope with two or more aquaria so that the average aquarist is unlikely to need to invest in the more robust, but expensive, piston pumps.

LIGHTING

Light is a most important consideration in the aquarium because it affects so many functions. The health of plants is directly linked to the amount, and more importantly, the duration of the light they receive. Lighting provides the energy plants need for photosynthesis. Fishes also require light; it is vital for their metabolism, and in many species it will influence their reproductive capacity. It will also affect their color as is easily appreciated if you consider the pallid fishes that are native to low light habitats— such as cave

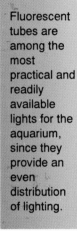

Fluorescent tubes are among the most practical and readily available lights for the aquarium, since they provide an even distribution of lighting.

Multiple fluorescent lights can greatly increase the illumination of an aquarium, an especially valuable feature for the deep or planted aquarium.

fishes and plants artificially denied light. In such cases they exhibit little or no color. A little research will reveal the kind of lighting with which your fishes are most comfortable. Some catfish, for example, are dwellers of dim places and can become wan and ill if kept in brightly lit aquaria. Discus are found in shadowy but clear waters where dappled sunlight plays all day. Likewise, livebearers often favor sunny aquaria.

There is no reason, however, to go to extremes with the lighting. For an aquarium with a depth of 20 inches two side-by-side fluorescent tubes are satisfactory. With a 24-inch deep aquarium, a third fluorescent tube can be added. Warm-tone lights are preferred. The color of light depends somewhat on the aquarium owner's taste. If you prefer more subdued light which gives the fishes a reddish tint, you should use the familiar Gro-Lux tubes. If you prefer more natural light, however, then the use of full-spectrum lights, which are commercially available from various firms, is recommended. Your aquarium shop owner will be able to help you select suitable tubes.

All lights will lose a percentage of their strength as they age. This loss can exceed 20%, and if considered in conjunction with even moderate cloudiness in the water, as well as increased

absorption of light by encrusted glass covers, you can see how the amount of light reaching the plants will slowly decrease with time. Keep this in mind and replace the lights periodically. Replacement about every 9 to 12 months is suggested.

The daily light period is 12 to 14 hours and should be regulated with a timer for the benefit of both fishes and plants. Plants can accommodate unnatural light conditions within reason as long as the duration of light is correct. In any case the light should be switched on a half hour before the first feeding. In this way the fishes can adjust to the new day and are already suitably active when first fed.

If the aquarium is located near a window, it will also receive a corresponding amount of daylight. This can mean that the aquarium lighting will not have to be turned on until later in the day, and can still be lit in the evening when most people can relax and enjoy their fishes. Aquaria should not be exposed to direct sunlight since they will be overgrown with algae and the water could overheat. Algae problems scarcely ever turn up with artificial lighting.

While fluorescent tube lights are excellent, if they have a drawback it is that their diffused light does not penetrate to the lower levels of very deep tanks, nor are they very good for creating variable lighting effects. For those wanting especially bright light,

Metal-halide lamps will provide a tremendous amount of light for the aquarium. However, they are quite a bit more expensive than fluorescent lighting.

The interior decorations (including the gravel) in an aquarium, should reflect one's own personal tastes.

other lamps are available. Metal-halide lamps (quartz-halogen) produce very bright light but are expensive. They will need to be suspended about 12 inches above the aquarium, which should have no canopy or hood.

Mercury vapor lamps are somewhat less expensive than the metal-halides, yet they generate considerable brightness, especially when housed in a suitable reflector. Again, they are suspended above an open-top aquarium.

Sodium vapor lamps are available, but since they have a strong bias toward the red end of the spectrum their application in aquatics is limited unless they are combined with blue-biased lighting.

There are many varieties of spotlights that can be used over aquaria, and by fitting them with various shades one can direct beams of light to specific parts of the aquarium to create stunning light-and-dark areas.

GRAVEL

There are probably as many kinds of gravel on the market as there are aquarium ornaments. For most applications, the natural colored medium pebbles are very desirable. Of course, if you want to buy gravel to match your drapes, it is available, and you are certainly free to do so. Usually about three inches of substrate is used. This works out to a pound or more of gravel per gallon capacity depending on the bottom dimension of your aquarium. The pebbles should be rounded and even in size. Many fishes will delight in taking mouthfuls and moving the gravel around and piling it up into little hills. For this reason you should never use glass particles as a substrate, for not only will it damage the fins and scales, but mouths as well.

USEFUL DECORATIONS

Rocks and pieces of slate are useful as decoratives in the aquarium. Rocks are handy to anchor floatables, build caves for fishes who use them, and to disguise equipment. While the world is full of rocks, free for the taking, do yourself and your fishes a favor and stick to safe rocks available at your pet shop. Unless you are absolutely certain the rocks you find in your back yard are not going to leach toxins into the water do not use them. The pet shop rocks are far cheaper than the cost of replacing a tankful of fishes.

DRIFTWOOD

Driftwood is a valuable addition to most aquarium designs. Many times you will be glad you have a piece of driftwood in the tank—not only because it looks good—but you'll find your fishes using it to get away from attentive mates or to define territories. The same caveat holds true as for rocks. Do not use driftwood you find on the beach or in your local lake. It's quite probable that it is full of unwanted passengers and/or their eggs, and you just don't know its past history.

THE BACKGROUND

The background is one accessory that may seem a little frivolous at first, but unless the aquarium is planned as a room divider, i.e., meant to be viewed

A variety of rocks used in conjunction with gravel and plants add to the visual impact of an aquarium.

from both sides, it really is more than mere decoration. When the rear and side glass of the aquarium are covered, the fish naturally feel more secure. Covering the back of the tank also gives you the opportunity to hide the cords, airlines, and other utilities. Last, but not least, the view through and through the tank is not the most desirable. Looking through clear to the wall behind the tank does not show your fishes off to their best advantage. Various attractive backgrounds are available from your pet dealer, but some people enjoy designing their own backgrounds. Another possibility is to paint the outside of the aquarium with waterproof paint. Synthetic enamel has proven to be a good choice. Do not choose a dark color; a light one is better. From experience it can be said that light green or light brown are optimal.

Fish are accustomed to receiving their light from above, and in the confines of the aquarium can be very nervous about light that comes from other directions. Since fish are phototrophic (they lean towards the light), it is not inconceivable that if your aquarium received substantial light from one side that the fish would eventually swim at an angle.

The insides of the rear and side glass can also be covered with a suitable decorative material. For example, pieces of slate can certainly be layered in the background or sheets of cork glued on. Decorative materials installed in the aquarium, however, must be tested for possible toxicity. It is best to stick with known commodities when putting anything into your aquarium. Any good pet shop will offer a vast array of *safe* decorative items for the aquarium.

Water of Life

Water is the only home your fishes will ever know. If the water is pure, clean, and of the proper temperature and chemistry, your fishes will likely live long and happy lives, provided their other basic needs are met. If, however, the water is too hot or too cold, or too acidic or too alkaline, or to hard or too soft, the lives of the fishes in both duration and quality are severely reduced.

WATER CHEMISTRY

Hardness is a way of expressing the amount of carbonates, bicarbonates, sulfates, and other salts dissolved in water, with calcium and magnesium being of primary concern. Hardness is expressed on the German scale as degrees of hardness (DH), which is a measurement of one part of calcium carbonate dissolved in 100,000 parts of water. Soft water usually has a DH of less than 3; medium water is about 6 DH; hard water is anything 12 DH or above.

Hardness of the aquarium water plays a subordinate role in the show tank. If the objective is the successful keeping, rather than breeding of your soft-water fishes, hardness is not a very large issue. Therefore, the water can by all means be medium-hard, which increases the stability of the pH. Apparently the biological stability of the aquarium system is more important than pH or hardness to the well-being of the fishes. A sudden drop or rise of pH in a short time, however, can be particularly damaging to fishes.

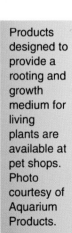

Products designed to provide a rooting and growth medium for living plants are available at pet shops. Photo courtesy of Aquarium Products.

Test kits for measuring water hardness are easy to use and widely available. Photo courtesy Aquarium Pharmaceuti-cals.

FRESH WATER
WATER HARDNESS TEST KIT

75 Tests

WATER HARDNESS TEST
BOTTLE NO. 1
SODIUM HYDROXIDE

WATER HARDNESS TEST
BOTTLE NO. 2

WATER HARDNESS TEST
BOTTLE NO. 3

When this happens it is termed pH shock and is extremely stressful to fishes. Be aware of this when you are adding pH-adjusting chemicals or moving fishes from one tank to another.

Rasbora heteromorpha comes from Southeast Asia and does best in soft, acidic water; highly acidic water is required for spawning success. Photo by M.P. and C. Piednoir.

The hardness and softness of your water is casually determined by how easily soap lathers for you. In hard water areas, more soap is required to produce a lather (and to get your clothes clean). Hardness is increased by the addition of special salts that will add minerals to soft water. Softness is achieved through the use of special procedures, i.e., reverse osmosis, distillation, and ionization. If you have a small aquarium and would like to soften your water a little, distilled water mixed with tap water is usually adequate. Generally, fishes that require soft water are happy enough if the pH alone is reduced to the recommended value without adjusting the hardness too much. It is easier to add minerals to the water, thus raising the hardness level, than it is to reduce the hardness of the water.

With pH 7.0 as neutral, lower values are in the acid range and greater values are in the alkaline range. There are a number of test kits available that will accurately measure the pH of your water. The tests are very inexpensive and, in my opinion, absolutely necessary for the proper care of your fishes. The test kits also contain chemicals that will lower or raise your pH to the desired values. While pH is often mentioned in conjunction with the softness or hardness of the water, generally the pH is lower in softer waters and higher in more alkaline waters.

Since the pH scale is logarithmic, the degrees on the pH scale mark a tenfold increase or decrease from the next higher or lower number, so a pH of 7 is vastly less alkaline than a pH of 8. This is not the area for approximation. An apparently small sudden change in pH is very stressful to most fish.

While most of the fishes available in pet shops are tank-raised, and therefore acclimated to somewhat different pH levels than those found in their native waters, it is in their best interests to make any changes in pH to optimal levels very slowly once you have established them in their new homes. Ask your fish dealer the pH under which they have been kept when you purchase your fish and bear in mind how stressful sudden changes in pH are to the fishes when you introduce them to your tank.

Once you have adjusted the water to the desired pH, it is kept fairly stable by the substrate and the aquatic plants. However, even the most alkaline water will slowly become more acidic as the fishes contribute their waste products (ammonia) to the water. This is why we always recommend partial water changes and monitoring of the pH levels in the aquarium. There are many products available that will lower and raise the pH of your water. When testing for pH, aerate the sample before testing and you will get more accurate results from your test kit. If all other conditions in the community aquarium are ideal, your fishes will still remain vital and healthy even if the pH is kept at approximately the preferred value for your fishes.

AMMONIA, NITRITES, AND NITRATES

Bacteria (*Nitrosomas*) in the filter medium is always a prerequisite for a properly functioning filter. The hobbyist must recognize that the filter medium does not clean the water and decompose toxins, but rather the bacteria perform this function. The biological filtration performed by these bacteria is critical for the prevention of high ammonia, nitrate, and nitrite levels in the aquarium. The action of

Products that set the pH to a pre-determined level adjust the chemistry of the aquarium water to suit a particular species. Photo courtesy of Aquarium Pharmaceuticals.

the beneficial bacteria is termed the "Nitrogen Cycle."

Ammonia, nitrites, and nitrates result from the breakdown of waste materials in the aquarium. These waste products include fish excreta, left-over food, dead fish, plants, and even snails. These chemical by-products are unavoidable in the presence of metabolism, but there are steps that must be taken to prevent them from overwhelming the fishes you are keeping. In an established aquarium, the beneficial bacteria reduce the ammonia and nitrites to less harmful nitrates that are removed by your partial water changes.

Again, the test kit for ammonia, nitrites, and nitrates is inexpensive and indispensable. The tests are easy to perform and will give you the added assurance that all is well within your aquarium—or—will let you know that drastic remedial action is urgently necessary. That test kit and your own nose are invaluable accessories when it comes to evaluating your water quality.

Chlorine and Chloramine

Municipal water supplies are almost always treated with chlorine or chloramine. These chemicals are necessary to make the water safe for us humans to drink, but they are very hard on our fishes.

Chlorine is easily removed from the water by letting it sit overnight. The chlorine gas simply escapes into the air, especially if you aerate the water.

Chloramine is created by mixing ammonia with chlorine and is very bad for fishes. It will not be released by aging the water and you must use one of the chloramine-removing preparations available. Your pet shop personnel will be able to tell you if your local water is treated with chlorine or chloramine.

Above: Since ammonia is one of the most harmful substances in aquarium water, it should be tested for frequently. Photo courtesy of Aquarium Pharmaceuticals.

Right: Levels of chlorine and chloramine should be monitored at all times. Photo courtesy of Aquarium Pharmaceuticals.

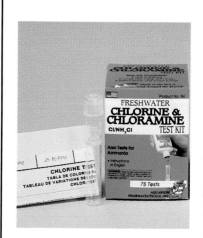

Aquarium Set-Up

An aquarium should not be placed in any area of the house where there is a great deal of traffic.

LOCATION

Like in real estate, when looking for a site to set up your aquarium, what matters is location, location, location. There are probably many places in your home where you could set up the aquarium, but some forethought now will certainly make life a lot simpler than discovering that you don't have enough room to pass comfortably when the aquarium is filled with water. There is no real "best" room for your new tank, but some thought should be given to the availability of electrical outlets, the water supply, and the fact that you won't want to move the aquarium after it has been set up. Kitchen, bathroom, bedroom, living room, den, or "fish room," your fishes could care less where the tank is located as long as their needs are met. Bear in mind that you don't want to place the tank too close to windows, heat or air conditioning vents, or in an area of the home that receives particularly heavy traffic.

Another important consideration, and one that probably wouldn't occur to a first-time aquarist, is the weight of the tank. Given that each gallon of water alone weighs over eight pounds, and that a good community size tank is 30 gallons, some thought should be given to the structural integrity of the floor. And this does not even include the weight of the tank, rocks, and interior equipment. If in doubt, check.

An aquarium should be situated in as convenient a place as possible to make servicing it less burdensome.

SET-UP OF THE TANK

You have purchased your tank and accessories and you are raring to go. You want to set up the most beautiful aquarium for your chosen fishes. First check and make sure you have a convenient water supply. The necessary water changes can become an onerous burden and are easily neglected if you have to haul water up and down stairs or from the far end of the house. The last thing you want to do is to scrimp on those water changes! Having selected an appropriate aquarium stand, place the stand in the chosen location. Do not place the stand too close to the wall. There will be many occasions when you will need to work at the back of the tank and you aren't going to be able to move it easily once it's filled with water. About 8 to 12 inches between the tank and the wall should be sufficient for maintenance.

It's best to use a soft pad, like styrofoam or felt, between the tank and the stand to compensate for any slight unevenness between the two that could possibly lead to a stress fracture in the glass or pop a seam. Even a piece of cardboard, cut to size, is suitable.

Rinse your tank with salt and fresh water. Fill it and check for leaks. If you do find a leak at this point, it is easily repaired with silicone. If you find one after the tank is set up, you will probably have to break the whole thing down again to find and reseal the leaking seam. Set the tank on the stand.

This is the time to make sure you like the arrangement. Is the stand even? Does the tank look "right" to your eye? Any changes you have to make are much easier at this point than after you have set the whole thing up. If I sound like I'm repeating myself, it's because at this point you can save yourself a lot of headaches if you savor the work at hand rather than rushing through to get to the finished product.

Don't be tempted to skip the gravel substrate in favor of cleanliness. (Many fishkeepers choose bare-bottomed tanks so that they can immediately

siphon off leftover food and fish wastes.) Some aquarists, in addition to fully covering the bottom of the tank with rocks and gravel, even paint three sides with an attractive, dark paint, leaving only the front glass bare to view the fishes. Most fishes feel very secure in this kind of cave-like aquarium.

Rocks are the decorative items of choice for most aquarists, but you will want at all costs to assure that there is no danger that your rocky arrangement will ever collapse. One way to minimize the danger of collapse is to select several of the largest, smoothest rocks available and cement them to the bottom of the

Decorations intended for use in aquariums should be made of materials that can be used without harming the fish. Photo courtesy of Blue Ribbon Pet Products.

tank with silicone. Let the silicone dry for 24 hours and you will have a good, solid base for further construction endeavors. It helps to draw a diagram of your planned arrangement prior to using the silicone because if you have to start over it will only delay operations even further.

You won't want to set the rocks directly on the glass, because if the glass is going to crack, it will do it where a rock is creating a pressure point.

Rinse, rinse, rinse your gravel. Even

Rocks also should be carefully washed before they are added to the aquarium.

after you have rinsed it enough, rinse it again. You will be amazed at how much sediment is carried in new gravel. When you fill your tank, you will see clouds of milky dust rising from the gravel. The only way to handle this is rinsing first and filtration later. You can rinse the gravel in a colander or a bucket, whichever method you prefer. Then gently pour the gravel into the tank. It is quite possible to crack your glass if you dump 30 pounds of wet gravel in all at once.

After you have your (optional) rock base and your gravel in the tank, you are then free to arrange the other rocks and driftwood to your liking, always bearing in mind that your fishes adore places with an entry and an exit. When you have to net a fish later, you will appreciate your forethought in providing exits to some of the caves. Put a net over the exit and drive the fish into the entry. Scoop up the net with the surprised fish. Sometimes it even works.

In lieu of rockwork, some aquarists are working with lengths of PVC pipe. While not quite as attractive as clever rock designs, there is a lot to be said for this method. You don't have to worry about the weight of the pipe, and it is quite simple to modify the decor. Just move it. If you chose you could create a "condo" effect with rows of PVC sections siliconed together. The fishes love it! In essence, if it's safe for the aquarium and makes a hollow space, most fish are going to use it. Also, when you are designing your tank, be sure to include some spaces that are too small for your adult fishes. Any fry you are lucky enough to get in the community aquarium will need plenty of havens for their first weeks until they

are large enough to safely swim out in the open.

Clay flowerpots sliced in half lengthwise are ideal for the aquarium. The material is inert, or non-toxic, the color is pleasing, and, best of all, they are very inexpensive.

After the tank is all set up with gravel, caves, plants, filter, and heater (but *not* turned on), and any other equipment you have chosen, add your water. Since you have already decorated your tank, use a gentle flow of water so you don't disturb your arrangement. Do use a chlorine/chloramine remover. Most municipal water supplies must treat the water to make it safe to drink, even if these chemicals aren't good for our fishes.

Once you have added your water, it is safe to turn on your heater and filter without worrying about cracking the glass on the heater or burning out parts of the filter.

Above: Using a bucket to transport water works great for small aquariums. On the other hand, a hose is more practical for filling larger aquariums.

Left: Gently pour in the new water so you don't disturb the gravel and decorations.

A properly set up and established aquarium will yield a tremendous amount of satisfaction for everyone.

Remember that it will take about 24 hours for the water temperature to stabilize, so do wait before buying your fishes.

When you have filled the aquarium with water, check the pH of your tap water. It is possible to check the pH of the water in a number of ways depending on how often you want to check the pH and how accurate you want to be. There are electronic pH meters on the market today where you just dip the electrode in the water and get an instant reading. These are useful if you have a large number of tanks to test and/or are very concerned about maintaining a stable pH level. The pH test kits that are available are simple and quite easy to use. Either way, when keeping fishes it is and will be necessary to know the pH of your water.

Acidic (low pH) water conditions can result from an excess of waste products in the water and are more likely to occur in an established aquarium. Brand new water is more likely to be closer to neutral (depending on where you live), at least until the fishes are introduced. But it is always possible that the water supply in your area provides soft, acid water or hard, alkaline water, so it is best to check.

CYCLING THE TANK

"Cycling" is a concept that is very familiar to marine aquarists and has nothing to do with bicycles. Basically, cycling a tank has to do with promoting the growth of beneficial bacteria (*Nitrobacter* and *Nitrosomonas*) in the aquarium. Without these bacteria the fish wastes and leftover food can turn a new aquarium into a toxic waste site in an amazingly short period of time. This condition in a newly set-up aquarium, manifested by almost milky, cloudy water, is known as "New Tank Syndrome." In an established aquarium, it is simply known as poor care.

The foundation of water quality is a process known as the "Nitrogen Cycle." Let's go back to the beneficial bacteria that converts ammonia into less harmful substances in the aquarium. Ammonia is the most toxic and the most common by-product of protein metabolism, i.e. fish waste. When combined with water of a high pH, the ammonia becomes about 10 times more toxic than it would be in the same amount in an aquarium with a neutral pH. Therefore, it is critical that ammonia be kept at

Once the aquarium water has cycled, it becomes stable enough to begin adding in new fish.

Kits that test for the nitrite level are available at your local pet shop. Photo courtesy of Aquarium Pharmaceuticals.

tolerable levels for your fish.

The nitrogen cycle describes the natural course of nitrogen metabolism in nature. Nitrogen is converted into ammonia as proteins break down and are acted upon by those beneficial *Nitrosomonas*. The bacterial action converts the toxic ammonia into nitrites, which in turn are acted upon by other bacteria, *Nitrobacter*, and converted into less toxic nitrates, which are then easily neutralized by plant (including algae) metabolism, zeolite and carbon in your filter, and your water changes.

In a newly established tank, the goal is to get a good head of steam going on the culture of these beneficial bacteria, so that when your tank is at fish load capacity, you will not have the dreaded ammonia problem. One way to accomplish this, when the tank is filled and the filter running, is to introduce bacterial cultures available at the pet shop. Then you can introduce one or two inexpensive, hardy fishes like small feeder guppies. Their wastes will feed the bacteria, which in turn will cause more bacteria to establish themselves in your filter. The filter is the home of these wonderful bacteria, and for this reason you should never clean your filter with hot water that will kill them off.

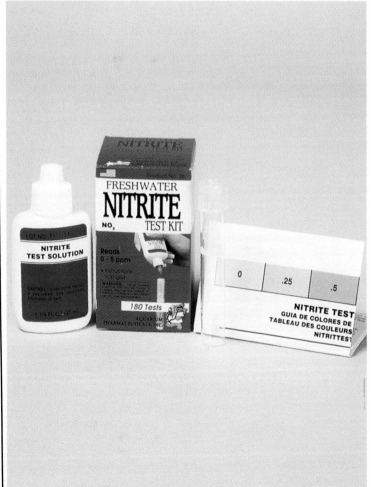

NEW TANK SYNDROME

When your new aquarium is first filled with water, except for suspended dust from the gravel, the water will be clear as it was when it came from the tap. Then you add the fish, and uh-oh, the water turns cloudy and milky looking. This phenomenon is known as "New Tank Syndrome" and is as benign as it is alarming. The cloudy water conditions are caused by surges in ammonia and nitrites in an aquarium that has not yet developed the beneficial bacterial colonies necessary to combat the surge. One way to avoid or lessen the clouding of the water is to only introduce one or two small fish in the first week or so of aquarium operation. Very few new hobbyists have the restraint to do this. Everyone wants to see their new aquarium fully set up and loaded with desirable fishes. Unfortunately, this can lead to early loss of the more delicate species as their gills and mucus membranes are eroded by the high ammonia levels in the new tank. The addition of some aged aquarium water from an established, healthy tank, or some gravel from such an aquarium will go a long way in developing the

For the first few months after an aquarium has cycled, it is best to slowly introduce new fish, such as this *Corydoras reticulatus*, so that the additional waste products given off will not overwhelm the biological equilibrium of the aquarium.

Young fish, such as these Gold Angelfish, *Pterophyllum scalare*, may be preferred to adult fish because they are better able to settle in and adapt to new surroundings.

bacterial colony desired. New water is just too "clean" to establish this colony quickly and usually needs a little help, either through the latter suggestion or the purchase and use of starter bacteria available at the pet shop. What you must do, however, when you notice this clouding, is refrain from changing the aquarium water. If you change the water at this point, the syndrome will continue over and over again. The water should clear within four to seven days if the tank is lightly stocked and you are not overfeeding the fish.

INTRODUCING THE FISHES

This is the moment we have all been waiting for! All your hard work is about to be rewarded. You have a good filtration system, the pH is steady, and the ammonia, nitrite, and nitrate levels are 0. Now it is time to introduce the fishes.

When you buy your fish, it is likely that they will be young. Most pet shops carry juvenile fishes. And this is good. It is better, especially when dealing with some of the more territorial fishes, to give them a chance to get adjusted to each other when they are still young and relatively docile. Take into account when you are stocking your tank that these little two-inch fishes will, in a short time, mature and double in size. Therefore, count on the tank looking a little bare in the beginning—but not to worry,—there are fishes there!

There are several ways of actually putting the fishes into the tank, but the one that I have found to work best is to put the fishes and the water they came in into a clean container. Gradually add an equal amount of water from the aquarium. Take your time with this procedure, giving the fishes time to acclimate to the

Pseudo-tropheus lombardoi is one of the species best purchased in small groups so that their aggressive tendencies will evenly spread out among themselves, thereby lessening their overall aggression towards other tankmates.

new water and then net the fishes out and put them into the aquarium. *Do not* put the water into your aquarium. Do not feed the fishes for the first day. They will probably be too interested in their new surroundings to eat and uneaten food combined with new fish waste can quickly cause water quality problems.

Leave the lights off for 24 hours to give everyone a chance to settle in. This will reduce petty squabbling and relax the fishes. You will see them soon.

QUARANTINE

Before new fishes are introduced into an existing aquarium, they should go through a suitable quarantine period to prevent possible introduction of diseases to the established community. I always feel it's best to be able to observe a new fish for seven to ten days before he is let loose on the tank, perhaps to sicken, hide, contaminate, and die. Often a health problem is easily cured in a small hospital tank, but the very devil to eradicate from the main tank. It is much easier to treat one fish in a small amount of water than all the fishes and the larger aquarium as well. It is also a good idea to make sure your new fish is eating as well.

On the whole it is advisable to introduce as few new fishes into the aquarium as possible. It is better to determine at the start which fishes will be stocked and to introduce them together within a short period of time.

Foods and Feeding

One of the highlights of fishkeeping is feeding them. This *Fossoro-chromis rostratus* is hand-tame and enjoys a meal from its owner. Most of the larger fishes will eat from your hand when they get to know you.

One of the greatest pleasures of keeping fishes is feeding them, and most fishes will not be picky about their food. Healthy fishes have good appetites. They are quite happy to eat just as often as you are willing to feed them. When you see a fish that isn't eating, most likely it is sick or harassed, or in some cases, tending fry.

Several small meals a day are more beneficial to your fishes than one or two large feedings. One or two large feedings encourages them to overeat which could, in turn, contribute to bloat and other digestive disorders. In nature fishes graze almost constantly so they are getting small quantities of food all day long. This is not usually practical for those of us who must spend time away from our aquaria during the day, but several small feedings will net us healthier fishes.

ON THE MENU

Herbivorous fishes are vegetarians and require a

Bloodworms make for an excellent treat for nearly all aquarium fish.

diet high in plant matter. Many kinds of African cichlids, the suckermouth catfishes, most livebearers, and many other fishes need vegetables to meet their nutritional requirements. You will notice that these fishes pick at everything in the aquarium. They are following their natural grazing instincts and for this reason are happier in an aquarium that has a good growth of algae.

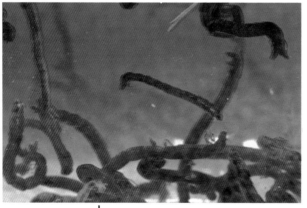

While brine shrimp is the most popular live food of choice, it should be offered sparingly, since too much may impact your fishes' intestinal tract.

In most aquaria they will quickly graze back whatever algae that naturally occurs. If you have a spare tank in the house, it is easy to culture some extra algae for their gustatorial pleasure. Set up a small aquarium with some fist-size rocks, add some aquarium water, plenty of light, and presto, instant fish food. When you get a good growth of algae on a rock, trade it for one in your aquarium and you will see your fishes immediately start to pick the algae off the new rock.

Other good greens for fishes include boiled spinach, crushed peas, frozen and re-thawed lettuce, and of course vegetable-based commercial foods. *Spirulina*-based foods are excellent. Frozen food blocks for marine fishes such as angels, tangs, and other marine vegetarians will also send them into a feeding frenzy.

Live Foods

Carnivores need a primarily meaty diet. Omnivores eat both meat and vegetable foods. While fishes are not fussy, for the meaty part of their diet live brine shrimp (*Artemia salina*) are tops. Perhaps part of the reason fishes are so fond of brine shrimp is that they dine on algae, giving them their meat and vegetable courses in one handy bite-size morsel.

Although live brine shrimp are first choice with any fish, for our convenience brine shrimp are also available in frozen and freeze-dried forms. The frozen brine shrimp do lose some of their nutritional value, but are still a valuable addition to your fishes' diet.

Bloodworms are rich; therefore, they should be fed only in small quantities as a treat. Tubificid worms are not recommended for fishes as they have been implicated in many bacterial infections. However, most fishes do love them, and if they are well rinsed, may be worth the risk, especially for desirable fish like clown loaches that can be difficult to start feeding. When you buy tubifex, check to make sure that the worms are actually alive. They should present as a round, wriggling, red ball. Grey and dead worms should be discarded.

Goldfish and guppies are

These *Corydoras imitator*, like almost all other freshwater tropical fish, will always be eager to accept tubificid worms in their diet. A specially designed worm feeder is an ideal device for holding them in place so that fish can easily dine on them.

Goldfish can make a great live food item for many large tropical fish.

live foods to your aquarium fishes. They are often thought of as the *haute cuisine* of the aquarium, but in truth they should really be considered the "meat and potatoes." Live foods give your fishes a zest for life that just can't be duplicated. Live foods are used by fish breeders for conditioning the prospective parents prior to spawning and many fishes simply will not breed without this food.

There is tremendous variety on the live food menu. Just imagine how many different kinds of insects there are in the world to get some idea of the kinds of live foods available. Many of these insects will send your fishes into a feeding frenzy. Just make very sure that when you do feed the fishes fresh insects that there is no possibility that they could have been contaminated with insecticides.

Many of the preferred live foods can be cultured right in your own home. Pet shops and live food specialists advertising in *Tropical Fish Hobbyist* magazine offer starter cultures for foods like *Daphnia*, white worms (*Enchytraeus*), and specially developed wingless fruit flies.

an excellent food source for larger fish, but feeding them live fishes will usually make them more aggressive. They get "used" to the idea of being predators and could easily "mistake" smaller tankmates for food. Live fishes are not necessary for their health or well-being, so encouraging them to be more aggressive is probably not a great idea.

We can never emphasize enough the value of feeding

Prepared Foods

There are many wonderful prepared foods on the market. There are specialized flakes, pellets, granules, freeze-dried live foods, and on and on. Often the dried foods even come with a feeder to measure out the food and keep it in one area of the tank. While each manufacturer claims his formula provides complete nutrition for your fishes, most experienced aquarists use several different brands in rotation to get the best results.

FROZEN FOODS

Most pet shops also carry a wide variety of frozen foods like brine shrimp, bloodworms, krill, etc. When you feed frozen foods, break a portion off the frozen block and put it in a fine net, then run some water over the block to 1) thaw the food, and 2) rinse off the water-fouling soup that surrounds the food. Then simply turn the net inside out in the tank and swish the food off the net. Your fishes will love it.

Your fishes will also relish tiny bits of shrimp, clam, mussel, crab, and fish. A small piece of frozen minced cod will last a long time. It's fun to give your fishes a variety of foods and they don't all have to come in a package. All it takes is a little imagination and your fishes can have a

Frozen foods offer a wide range of choices to the aquarist. Photo courtesy of Ocean Nutrition.

different dinner every night of the week.

PORTION SIZE

Frequency of feeding is often determined by our schedules rather than our desires. Fishes will *survive* quite well with one feeding per day. Fishes will thrive and grow well with two, three, or more feedings per day. The trick is never to overfeed the tank. A good rule of thumb is to feed no more than your fishes will eat in five to fifteen minutes.

Observe the fishes and the character of the foods you are offering. If the food floats slowly to the bottom the fishes are going to be able to eat more of it than heavy food that falls immediately to the bottom and into the spaces between the gravel, fouling the water and making a mess. Some foods take a very long time before they sink and you can soon decide just how much is enough for your fishes. Start with a portion that will fit on a dime and see if they eat all of it within the five minutes allotted. If they scoff it up in two minutes, double the ration. If you notice that they are sated and there's food to spare, simply reduce the ration.

FASTING

Just as it is healthful for most people to fast for a day, it is also good for your fishes. Fasting will give their digestive systems a rest and promote a good appetite the following day.

It is important to begin feeding your new fish sparingly (right) until you get a feel for just how much they can eat at any given time.

Flake foods, available in a variety of conveniently sized packages, are among the most commomly offered fish foods. Photo courtesy of Wardley.

These Blue Platies, *Xiphophorus maculatus*, will thrive on a combined diet of live, vegetable, and flake foods.

Maintenance

A well-aged system will house more fishes than a newly established aquarium, but common sense and restraint are important even when keeping such hardy fish as this *Macropodus opercularis*.

The true aquarist is revealed in the restraint shown in stocking the aquarium. A restricted biologic circulation prevails in the aquarium, which is easily disturbed by overpopulation with fishes.

HOW MANY FISH?

The question of how many fishes can be kept in an aquarium has been asked since the first Paradisefish (*Macropodus opercularis*) was imported to France about 100 years ago, thus starting the delightful tropical fishkeeping hobby. There are many variables to be taken into account when stocking the tank. First in my mind is the maturity of the system. If you have a brand spanking new aquarium without a hint of biological activity, the recommended numbers don't apply—yet. To stock an immature system to its optimal level is simply courting disaster. After the water and filter have aged a bit, then you can slowly add new fishes, but in the first weeks keep the number of fishes low and you will have a far better chance of long-term success.

In truth, there is no hard and fast rule for how many fishes your aquarium can accommodate. The old rule was one inch of fish per gallon of water, but unfortunately most people took that rule too literally and many fish suffered needlessly. Imagine, if you can, two 10-inch Oscars in a 20-gallon tank! Twenty one-inch Neon Tetras, however, pose no problem in the same 20 gallons of water. Common sense dictates that the number of fishes you can maintain depends on their adult size, personalities, the

The one inch of fish per one gallon of water rule applies primarily to small, relatively non-aggressive fishes. These Neon Tetras, *Para-cheirodon innesi*, for example, are happiest in a crowd.

efficiency of your equipment, and your attention to water changes, feeding practices, etc. The shape of your tank is also an important factor in the stocking density of your aquarium. A narrow, deep tank will hold far less fishes than a low, wide tank. The surface dimension is almost as important in determining fish capacity as the number of gallons of water in the tank. It is at the surface that the very important exchange of carbon dioxide for oxygen takes place and without adequate oxygen in the water, no fish can survive.

TENDING THE TANK

Once you understand the value of water changes, there really isn't a lot more to do than enjoy your fishes. Of course, the filter medium must be changed every so often, usually about once or twice a month depending on your fish load. Algae can be scraped off the front glass and plants trimmed, but constant fussing is neither desirable nor necessary.

If you understand the

The filter medium in this box filter is at the point where it needs to be changed or rinsed out with luke warm - water.

filter medium should only be rinsed briefly in lukewarm water. This is particularly true of filter systems equipped with a sponge.

These filters are primarily used to encourage the nitrogen cycle to function. The media provide a home for the beneficial bacteria to colonize. The most obvious medium is the gravel substrate, but there are many other possibilities, including the medium in most mechanical filters.

The surface of the filter media is ideal for the colonization of these beneficial bacteria, and usually contain a ready source of food in the form of trapped organic matter. Biological filters must of course have a good air supply and they do take some months to build-up the effective colony size. They are easily destroyed by medicines added to the aquarium, so you must always be careful on this account—some medications claim not to harm such colonies.

If the bacterial colony is destroyed as with over-zealous cleaning with hot water or if the power is off for some hours the water will turn sour and give the owner the impression that the problem is lack of

effect of bacterial action on filtration, it makes absolutely no sense to clean the filter in such a way that the entire filter medium is changed or scalded with boiling water. This suddenly removes all beneficial bacteria and the fresh filter medium cannot assume the function of cleaning the water in the aquarium because it is completely inert. Such a mishandled aquarium could even "lose its balance." It takes several days or even weeks for new bacteria to become re-established in replenished filters. Only then will the filter be biologically active again. For this reason the oxygen, as the signs are much the same—the fishes gasp at the surface.

When your aquarium is properly set up, maintenance is a relatively simple matter. Basically, all you need to do is change about 25 percent of the water every two weeks, vacuum or stir the gravel, and service the filter(s). But of course, that is a bare bones maintenance and will not do much to enhance the beauty of your show tank.

There are many devices on the market to make your water changing chores easier. The only advice is that you find a system that you are comfortable with and keep up those water changes! The amount and frequency of your water changes are largely dependent upon the number of fishes you are keeping, how much you are feeding them, and your ammonia test readings. You will soon get a feel for when a partial water change should be done. Food that remains lying on the bottom is a problem for the water quality of an aquarium. In smaller show tanks, in particular, you must make sure that these food remains are siphoned off without fail. Small, frequent water changes are most appreciated by your fishes.

Non-submersible type heaters are as commonly used as submersible types. Photo courtesy of Hagen.

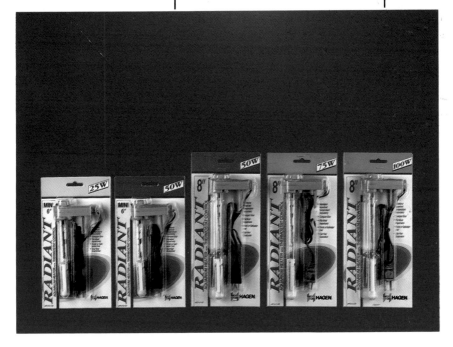

WATER QUALITY

The *most* important aspect of water quality is cleanliness. All the wastes produced by your fishes, plus any decaying plant material, plus uneaten food, dead fishes, etc., all add to the pollution level of your tank. Fortunately, fishes are very hardy and can withstand quite a bit of neglect (except for those

fishes relative to the size of your tank, not overfeeding, having an adequate filtration system, using carbon and zeolite in addition to your filter media, and doing frequent partial water changes.

If you have been lax about water changes, be very careful when you do start changing the water again. Your fishes have

It cannot be stressed enough how important it is to perform partial water changes on a regular basis in order to maintain the well-being and overall health of the aquarium inhabitants.

high ammonia levels) before they show signs of stress. But we assume you want your fishes to do more than just survive, and in this case it is important to keep the dissolved wastes in the tank to a minimum.

There are several things that contribute to the cleanliness of your aquarium water. These include keeping a reasonable population of

become accustomed to the high pollution level in the tank and if you try to mend your ways by doing a massive water change all at once, you could very well kill your fishes with kindness. If you can't remember when you did your last water change, take it easy. Start by doing small water changes daily for a couple of weeks. If the fishes are tolerating this well, start to

increase the amount of water changed, and decrease the frequency of changes until you feel the fishes are looking their best and the water parameters are where you want them to be.

How often and how much water you change depends on many variables: how many fishes there are, the type and amount of food offered, the temperature, etc. Just remember that small, frequent water changes are less stressful to your fishes than a large water change all at once. A 10% water change once a week is not too much, and is probably a small enough amount of water not to be too much work. If you have a heavy fish load in your tank you may need to change more water than that.

Water changes have an invigorating effect on your fishes. They will often trigger a reluctant, but almost-ready pair to spawn and the nice clean water will help ensure an optimal hatch rate.

When you do your water changes, be sure to match the water temp-erature to within a few degrees, and add a

Above: Fish showing their best colors with erect fins, as seen in these discus, *Symphysodon spp.*, are a good sign that the conditions in your aquarium are fine.

Left: These sensitive Clown Loaches, *Botia macracantha*, may require more frequent, smaller water changes.

For those aquarists who have difficulty maintaining clear water conditions, a diatomaceous earth filter will clear up the water within a couple of hours.

product that neutralizes chlorine and/or chloramine. Check your pH and add appropriate buffers to keep the pH level stable. Remember, do not make any sudden changes in pH.

Bear in mind that evaporation of the water in your tank will affect pH and ammonia levels. As the water evaporates, the minerals, salts, and ammonia, nitrates, and nitrites will remain in the water in greater concentrations, so it is important if you are not doing a water change to top off the tank regularly to dilute these elements.

GAS EMBOLISM

In cold climates, there is a serious problem that can occur if you replace tank water with water straight from the tap. It is called *gas embolism*. You can lose every fish in your tank to this condition. Cold water has an enhanced capacity to hold dissolved gases such as chlorine, oxygen, and nitrogen. In the summer, the tepid tap water is relatively benign except for chlorine and chloramine. In the winter, this water becomes lethal with gas. When the gas-loaded water warms up in your tank (or from the addition of some warm water to bring it to tank temperature), the gas appears in the form of bubbles that cover every surface in the tank—including the fishes. The fish's gills are quite permeable to the dissolved gases and bubbles will form in the capillaries of the fishes, leading to intense pain and horrible hemorrhages. The prophylaxis is very simple—age the water for 24 hours and the gas will escape harmlessly.

DIATOMACEOUS FILTER

A diatomaceous earth filter is an invaluable piece of equipment for people who have problems with clouding of the water, whether it be from an algal bloom, bacterial bloom, or overfeeding. In essence, the diatomaceous earth filter utilizes diatomaceous earth as a filter media, straining out microscopic particles from the water and effectively "polishing" your water. A couple of hours of diatomaceous earth filtration will render your aquarium water crystal clear!

BLOOMING ALGAE

Blue-green algae is a common problem in aquaria. When keeping fishes, you want to encourage the growth of green algae, but the blue-green kind won't impress anyone, not even the fishes. Blue-green algae is almost always a sign of anaerobic conditions in the aquarium. It often appears bright green and "greasy." It gives off a "fishy" odor. Anaerobic conditions are the result of decreased oxygenation and are commonly found in areas of the aquarium that has little or no flow of water.

Organic material gets trapped in between the grains of gravel, where it will decay, preventing the water from circulating through the gravel. The lack of circulation results in areas of the substrate where oxygen levels are low or absent. This condition will eventually become toxic to the fishes. It is therefore recommended that "gravel vacuuming" become part of the routine maintenance of every aquarium.

GRAVEL VACUUM

Changing water alone is not enough to keep your aquarium clean. You will be amazed at how much dirt (detritus) can collect in the substrate. If you were to take a stick and stir the gravel in an established aquarium, you probably wouldn't be able to see the fishes! Every bit of leftover food and fish waste will find its way into the crevices between the stones to decompose at leisure. There is a certain amount of

Algae can easily be removed from the aquarium glass through the use of magnetic algae scrapers. Photo courtesy of Hagen.

bacterial activity going on in the substrate to help break down this waste, but it's never enough. You will have to do your best to maintain a healthy substrate. This includes vacuuming the gravel with a device called a "gravel washer," a long, thick tube that literally sucks the waste out of the gravel. These devices are great for water changes. There are even devices that hook right up to your faucet, clean the gravel, remove the old water, drain the waste water down the sink, and add the new water right from the tap.

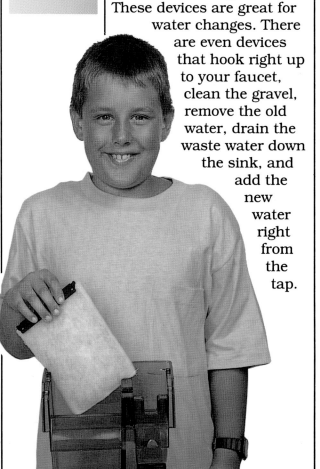

FILTER MEDIA

Your choice of filter will, in large part, determine which type of filter media you will use. Whether it be floss and charcoal, or filter pads, or hair rollers, you will need to follow the manufacturers recommendations about the maintenance of your filter. It is best, however, to: 1. Clean your filter only with lukewarm water (not too thoroughly) to keep your bacterial cultures going. 2. Clean your filter and change your media a few days after you do a water change and gravel wash. The established bacteria in the filter are best able to accommodate the extra load of stirred up wastes and a clean filter is less effective at this time. Besides, if you want to get the last bit of use out of your media, change it after it has trapped the extra waste.

The Fishes

Xiphophorus maculatus, the Platy, is commonly found in nearly every tropical fish store. This is a hardy fish that is beautiful as well, and highly recommended for beginners.

Freshwater tropical fish are one of the most beautiful and diverse groups of fishes available to the hobbyist. There are about 50 so-called "bread and butter" fishes that are available in almost every pet shop in the country. These fishes are the mainstay of the hobby in terms of beauty, relative ease of keeping, and popularity. In addition to the bread and butter fishes, each shop has its own selection of specialty fishes that are usually a little more exotic or require more specific care, i.e., special foods, water conditions, etc.

LIVEBEARERS

Nearly every new aquarist starts off with some livebearers and with good reason. The livebearing fishes of the family Poeciliidae breed naturally in the aquarium without any special attention, as long as they are well fed and cared for. Guppies, Platies, Mollies, and Swordtails are often called the "Big Four" and are the most popular of all the livebearers. They are hardy, peaceful, easy to care for, and remain a reasonable size so that it is entirely possible to have a good number of fishes in a modest-sized tank.

This captive-bred variety of *Poecilia reticulata*, known as a variegated snakeskin Guppy, sports a much larger and ornate tail fin than his wild relatives.

GUPPY
Poecilia reticulata

The Guppy is a prolific breeder as evidenced by its former common name, "Millions Fish." Guppies attain a size of no more than 1½ to 2½ inches with the females being larger than the males. Native to Trinidad, Venezuela, Barbados, and parts of northern Brazil and the Guianas, the Guppy is peaceful and very active. The original wild Guppy was a pretty little thing, about an inch to two inches long, at the most. The males were spotted red, yellow, and blue. The females were a drab grey. Guppy breeders, however, have bred for success. The "fancy" Guppies found in the pet shops today are hardly recognizable when compared to the wild stock. Fancy Guppies show up in some fantastic colors, sizes, and fin shapes.

Keep Guppies in neutral to slightly alkaline water, medium-hard water in the 60 to 90°F temperature range with about 75°F being ideal for the home aquarium. While Guppies will accept a wide variety of dried and live foods, their regular diet should alternate between quality general-purpose flake foods and vegetarian flake foods with live foods offered at

Most of the livebearers are found in tropical areas inhabiting coastal marshes and streams. Their native waters tend to be a bit on the hard, alkaline, and even brackish side, which makes the addition of about one tablespoon of rock or sea salt per five gallons of tank water a prerequisite for their good health. Mollies especially thrive when kept in brackish water.

All of the livebearers appreciate vegetable-based diets in addition to some wriggly live foods. A good quality flake food for vegetarians will provide a good staple diet.

breeding time.

Many high-class strains of fancy Guppies are available, but for breeding purposes, it is best not to mix and match colors and types indiscriminately. Mixing strains often results in the desirable traits of both parents being lost in the offspring.

PLATIES
Xiphophorus maculatus and *X. variatus*

The Platies are natives of Mexico. Colorful, attractive fish, Platies have been bred for so many color varieties and hybridized forms that at the present it is just as difficult to find the pure wild-type originals as it is to find a wild-type Guppy.

They are very hardy fish, and the females make up for their relative drabness by producing a great number of fry at regular intervals. The males are about two inches and the females about three. The fry are very easily raised, and the males do not show any color until maturity.

Keep Platies in a sunny aquarium with clean water in the 70 to 80°F range. Neutral, fairly hard water is best, but not essential.

Feed your Platies on flakes, live foods, and vegetable-based foods.

SAILFIN MOLLIES
Poecilia latipinna and *P. velifera*

Sailfin Mollies are among the loveliest of the livebearers. They are the largest, measuring four to five inches with a spectacular "sailfin" dorsal

The colors seen in the *Xiphophorus variatus* (above) and *Xiphophorus maculatus* (below) have been artifically produced in captivity.

These beautiful sailfin mollies, *Poecilia velifera* (right) and *Poecilia latipinna* (below), are among the largest livebearers available in the hobby.

fin. Sailfin Mollies are available in green, orange, brown, white, silvery, and black, and shades of each. The male alone sports the magnificent dorsal fin. They are natives of the southeastern coastal United States south to the Yucatan Peninsula.

Of all the readily available livebearers, the Sailfin Mollies are the most difficult to keep. They absolutely require brackish water conditions with one tablespoon of salt per gallon of water and vegetable matter in their diets. Keep the temperature steady at 75 to 80°F. The water should be neutral to slightly alkaline. Sailfin Mollies need a large tank with well-aerated water—20 gallons is the minimum tank size.

The females should not be moved close to their delivery date, but if well fed will not cannibalize their babies. Do not use a breeding trap for these fish.

The babies will grow quickly and if well fed and given ample space, will grow huge dorsal fins.

SPHENOPS MOLLY
Poecilia sphenops

The Sphenops Molly is a placid, peaceful fish that does well in the community tank with other fishes that can tolerate brackish water. It is a native of Mexico and Central America, and also parts of northern South America.

The Sphenops Molly grows to about four inches in length and is velvety black, mottled, black with orange in the tail, or gray-green with orange. Sphenops Mollies have hearty appetites and a great need for vegetables in their diets. Maintain the water at 55 to 80°F, adding one tablespoon of salt per gallon of water, keeping the water well filtered and aerated. The addition of live plants can only benefit Mollies.

Mollies are easy to feed, accepting flake foods, but thrive and breed best on a combination diet of flake, vegetable matter, and live foods.

SWORDTAIL
Xiphophorus helleri

The male Swordtail, as the name implies, sports an extension on his tail in the shape of a sword. Natives of southern Mexico to Guatemala, Swordtails are available in many colors and with varying sword sizes. The Swordtail measures about four to five inches, are hardy and long-

Like all other mollies, this *Poecilia shenops* prefers a small amount of salt in its water.

lived. They do, however, have a propensity for leaping, so be sure you cover the aquarium well.

They are active, busy fish, usually peaceful, but occasionally a bully male will show up in the community. Keep them in 75°F water, slightly brackish, and neutral to slightly alkaline.

They accept all regular aquarium foods, but there should be variety in the diet and additional greens.

Swordtails breed with extreme ease. Beginners have no difficulty sexing and breeding this species, but they sometimes have difficulty raising good-sized young. The basic requirements for optimal growth of fry are plenty of good, clean water and good food. Large females can give birth to 150 fry at a time.

BREEDING LIVEBEARERS

The differentiation of sexes is easy, fortunately, in most livebearers. The males characteristically possess a gonopodium, an organ of copulation that is formed from the modified anal fin. There is often a large difference in the size of the sexes, with the males being the smaller. This is particularly obvious in the Guppy in which the male is also colored and may have long and exquisitely beautiful fins, while the female is larger and less colorful.

Young livebearing females can be fertilized at a very early age if mature males are present, in the case of the Platies some eight days or so after birth. Even when they are fertilized very early, the females do not bear young for many weeks, and

The original wild swordtail does not have an entirely red body. Seen here is a pair of wild-caught swordtails, *Xiphophorus helleri.*

Guppies, Platies, and Swordtails may be expected to drop their first brood not younger than 10 to 12 weeks.

Males take no notice of whether a female is already gravid or even about to drop young, but pay court to all and sundry, even immature males. They hover around the female or chase her around the tank, often with a spreading of fins and, particularly in the swordtails, a backward swimming motion that is very characteristic. The female seems indifferent to all of this, and the male simply darts in and ejects his sperm when the chance presents itself. The sperm are stored in the female reproductive tract and fertilize successive crops of eggs for the next five or six months. If fertilization continues to occur, as it does in a mixed tank, the new sperm certainly fertilize some of the eggs, but the extent to which the original insemination can be superseded by later ones has never been fully worked out.

In the Guppy, Platy, Swordtail, and Molly, successive crops of eggs occur in batches so that one lot of young all the same age is produced followed about a month later by another batch. At the average temperature of about 75°F, the actual development of the young from the time of fertilization to birth takes about 24 days, while the brood

A great deal of close body contact is typical of many livebearing aquarium fishes. This is certainly the case with these *Poecilia velifera*.

Once livebearing young have been born, they should be separated from the parents and raised by themselves.

interval is about 30 days. The extra week is taken up by the development of the next crop of eggs prior to their actually being fertilized. Most livebearers produce young at about 22-day intervals when kept near to 81°F. and in bright light.

The embryos developing within the female poeciliid possess a yolk sac, a bag containing nourishment present in the egg when fertilization takes place. During the embryonic development the food stored within it is used up. The young fishes develop in a folded position, head to tail, and are often born with this fold still present. They may sink to the bottom for a short period but are usually able to fend for themselves either immediately or within a couple of hours. The average length for a newly born Swordtail or Platy would be about about one-quarter inch. The number of fry ranges from 30 or 40 for Mollies and 60 to 150 for Swordtails, Guppies, and Platies.

The pregnant female swells unmistakably in most livebearers and also exhibits the gravid spot, which is a dark spot near the base of the anal fin that is caused by stretching of the peritoneal wall. Moving the mother, particularly mollies, is likely to cause premature birth. She is best moved either early or very late, so that the young are in no danger when birth occurs. The young swim toward the light and if the tank is heavily stocked with fine-leaved plants, they will swarm into them and be safe from cannibalism.

Breeding traps have been used with varying degrees of success. They do not often suit the mothers or the fry. The mothers dash around and injure themselves and the water quality is usually not too good for the fry. It is more frequently practiced to use either a tank divider in a dark corner of the tank through which the young will escape to the light, or to suspend a cage in the community tank.

However, most breeders prefer the more natural method of transferring the mother to a small aquarium with abundant plants that provide shelter for the young and a used sponge filter to maintain good water quality. The mother is removed as soon as it is evident that she has dropped her fry, and the young are raised in the same tank. If the mother is well fed, particularly on live foods, it is likely that she won't cannibalize her young at all. Mollies will not even eat their young at all unless they are very hungry, so the best precaution when breeding mollies is simply to keep the parents well fed.

Livebearing young are quite large and can be fed crushed flake or other prepared foods right away. Feed live foods such as newly hatched brine shrimp, microworms, and other small foods for optimal growth.

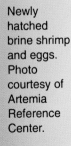

A divider can often be employed to separate mother from young (above). After the young become free-swimming, they can be fed live baby brine shrimp immediately.

Newly hatched brine shrimp and eggs. Photo courtesy of Artemia Reference Center.

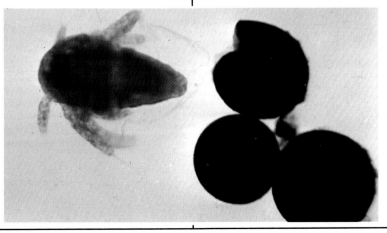

BARBS

Barbs are all members of the family Cyprinidae which includes some temperate fishes, i.e., Koi and Goldfish, etc. Tropical cyprinids are peaceful, attractive, and easy to maintain in the home aquarium.

ZEBRA DANIO
Brachydanio rerio

Zebras are one of the first fishes every new hobbyist falls in love with...and with good reason. They are active, cheerful bodies that are undemanding and delightful to watch as they school around the aquarium.

They are so busy that they do best in a long tank, but be sure it is well covered as when the walls confine them they can be apt to leap. Be sure to keep at least six of these small, 1½ - to 2- inch, gregarious fish together for their peace of mind. They are definitely schooling fish and get a bit spooky if kept without their own companions.

While they will tolerate a variety of water conditions, neutral to slightly acid and slightly soft water suits them best. Keep the water in the 70 to 85°F. range for optimal health. They will eat almost any aquarium food, but do enjoy a meal of live food now and then.

WHITE CLOUD
Tanichthys albonubes

The White Cloud is a native of the mountains of China where it makes its home in cold, fast-moving waters. They are very hardy and will breed at the drop of a hat. It is almost impossible to do injury to these wonderful little, 1- to

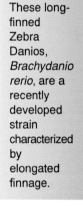

These long-finned Zebra Danios, *Brachydanio rerio*, are a recently developed strain characterized by elongated finnage.

1½-inch fish. They will accept almost any reasonable water conditions, but prefer neutral to slightly acid and soft to medium water. Keep the temperature anywhere from 40 to 78°F but don't make any sudden changes in temperature. Feed them well on prepared and live foods and they will live long and fruitful lives.

TIGER BARB
Barbus tetrazona

The Tiger Barb has to be one of the all-time favorite aquarium fishes. These 2- to 3-inch fish are very active and peaceful although they may not be able to resist nipping the fins of slow, elegant fish like the Betta. This nippiness is often minimized if the little tigers are kept in schools of 6 or 7 fish. Their time is better spent chasing each other around.

They are easy to feed. Always hungry, tigers will eat any aquarium food, but benefit from some vegetable matter in the diet.

Keep them in soft, slightly acid, warm water in the 78°F range and you will enjoy them for a long time.

They will breed readily in the aquarium in the manner of egg scattering fishes, but some accommodations must be made for saving the eggs or they will be quickly eaten.

Above: White Clouds are not tropical, but come from cool waters.

Below: Tiger Barbs need warm water to maintain health.

Corydoras aeneus is one of the most peaceful, common, and inexpensive catfishes for the aquarium.

CATFISHES

Catfishes are found all over the world and one common feature is that none of them have scales. They range in size from the diminutive *Otocinclus* and *Corydoras* species to the seventeen-foot-long *Silurus glanis*.

As a rule, the catfishes are hardy and adaptable. Most are nocturnal and do best in subdued light.

BRONZE CATFISH
Corydoras aeneus

C. aeneus are probably the most popular and best known of the armored catfish. Their coloration is pleasing but by no means flashy. Their body is a yellowish to a greenish brown with a metallic glint on the sides while their belly is a grayish yellow. These fish are found in shallow, muddy waters, frequently in shoals, all over a good part of South America, but most come from Trinidad.

They are quite peaceful. Since they spend so much of their time on the bottom of the tank nuzzling the gravel for food, they have been thought of as scavengers. Yes, they are useful in finding food in the substrate, but should be offered food suited to their needs as well. They need foods that will fall readily to the bottom since they only surface reluctantly.

They may reach a size of three inches but usually are somewhat smaller in captivity. They accept any aquarium food, dried or prepared, but they love tubifex worms above all other food.

Their water should be fresh and neutral to slightly alkaline. A hardness of around 10 DH is best since calcium is needed for their normal growth. They are native to waters of low oxygen content and have a form of intestinal respiration which helps to supplement the normal oxygen intake. They can be seen taking occasional gulps of air and then diving back down to the bottom. The intestines are lined with numerous tiny blood vessels and act much as a lung in osmotic respiratory exchange.

These catfish are alert and often comical, always in constant motion, actively surveying the bottom. They have a double pair of barbels which is useful in the dark in getting about safely and locating food. These also serve as accessory taste organs and play a part in mating. They are well protected with bony armor plates and so have little to fear, even from much larger fishes.

PEPPERED CORYDORAS
Corydoras paleatus

These catfish are found in southern Brazil and parts of northern Argentina in shallow muddy water. They are peaceful and are commonly used in many tanks. They reach a size of about three inches and begin to breed at two and

Since these Peppered Corydoras, *Corydoras paleatus*, are bottom-dwelling fish, food offered should always make its way down to the lower reaches of the aquarium.

Every pleco belonging to the genus *Hypostomus* will require some vegetable matter in its diet, especially larger plecos like this *Hypostomus scaphyceps.*

one-half inches. Feeding is no problem at all since they will eat all aquarium foods and do especially well on worms.

The Peppered Corydoras, as they are commonly called, do quite well in water which is neutral and rather soft, with a temperature range of 74 to 80°F.

They have dark olive brown to yellowish green flanks which have a metallic glint. They have mottled dark brown patches on a lighter background and spots on the dorsal and caudal fins. An albino variety exists.

They breed quite freely and will often spawn in a community tank.

PLECOS
Hypostomus spp.

There are many different species of *Hypostomus* and other South American suckermouthed catfishes.

Several species are strikingly beautiful while others are beautiful in their ugliness and are very useful in the aquarium for the removal of algae. They are primarily vegetarians, so you must offer them plant matter in addition to normal aquarium foods. Parboiled zucchini is a particular favorite. They can grow quite large but will remain stunted in smaller aquaria. Plecos are natives of the Amazon Basin and as such prefer neutral to slightly acid water that can be soft to medium hard.

Called Loricarids, suckermouth-type catfishes should be provided with driftwood in the aquarium, not only as a shelter during the day, but they like to rasp the wood with their mouths and it is thought that this is necessary for their digestion.

Apisto-gramma agassizi is an ideal dwarf cichlid for any peaceful community aquarium.

CICHLIDS

No single group of aquarium fishes offers the diversity of size, behavior, and habits as cichlids. They are fascinating in their breeding habits. They take care of their eggs and young. The extent of this care differs by species, but linked with this care is a strong tendency to savage other fishes of the same or different species. An established pair of cichlids may be left to their own devices and sooner or later they will probably spawn. When trying new pairs, the aquarist must remain watchful and be prepared to remove one or the other fish if trouble develops. If spawning has taken place and anything goes wrong, or if all of the young are suddenly removed, the pair may quarrel, as if they suspected each other of disposing of the young.

APISTO
Apistogramma agassizi

Apistos are dwarf cichlids from the middle Amazon region that are peaceful (except when spawning) and very suitable for a community aquarium. The average male grows to about three inches; his mate is about two inches. They will begin to breed when they are about two-thirds grown. They are not too difficult to sex. The female's fins are shorter, rounder and a more uniform color. The aquarium should be set up

Hemichromis lifalili, one of the jewel cichlids from Africa, attains the bright red coloration only during spawning activity.

in the manner suitable for most dwarf cichlids, i.e. rocks, caves, and plants.

A. agassizi are heat-lovers. The temperature may range from 80 to 85°F but an average of 82°F is best. Provide water which is well-aged and slightly acid. They will thrive on a variety of prepared foods, which can be alternated occasionally with frozen and live.

JEWEL CICHLID
Hemichromis lifalili

Jewel Cichlids are found throughout tropical Africa. They are very pugnacious and have the unfortunate habit of uprooting plants. They are fairly hardy and extremely colorful, especially at breeding time, and if it weren't for their nasty dispositions they would be one of the most

popular aquarium fishes.

They may reach a size of six inches but average about four. Although they prefer live food, freeze-dried foods will be accepted. The water requirements are not stringent although they prefer warm temperatures between 78 and 85°F.

They have beautiful coloration, especially at breeding time. Their sides are greenish yellow, sprinkled with large brilliant green flecks. The fins are a yellowish brown to a greenish, the tail is edged with red. Both sexes undergo a magnificent change in color when they are about to breed. The yellowish green of the body becomes an intense red color. The female is even more brilliant than her mate, which is unique among cichlids.

The tank should be provided with numerous hiding places. A flowerpot, several rocks, and plant thickets provide them with the privacy they require. They spawn in the typical cichlid fashion and are very conscientious parents. The fry are small when they hatch and should be fed powdered fry food and live baby brine shrimp at first and later small daphnia, crushed flakes and freeze-dried foods.

FIREMOUTH CICHLID
Herichthys meeki

Firemouths are beautiful and very popular fish. They are rather peaceful for cichlids but cannot be kept with smaller fishes. They are carnivorous and should be given a varied diet of predominantly freeze-dried, frozen, or live food. They can become quite tame and learn to take food from your fingers. If fed a proper diet the male will reach a size of five inches while the female is about four inches. When in top condition they will breed at two inches. The water conditions are not critical, but extremes should be avoided. They do, however, prefer that the water be slightly alkaline (7.2 pH). The temperature may range from 74 to 85°F. It is necessary to keep them in a large tank, 30 gallons or greater, with some open areas and a number of rock formations for shelter.

Although a relatively peaceful cichlid, the Firemouth, *Herichthys meeki*, should not be trusted with fish significantly smaller than itself.

C. meeki are native to northern Yucatan. They have quite flashy coloration. Their body is a bluish gray, with a violet sheen and their back is darker. There are a number of vertical bars and a horizontal stripe runs across the body. The unpaired fins are edged with blue-green while the other fins are a yellowish to reddish brown. A fiery reddish-orange covers the belly and sometimes runs up into the throat and mouth.

Members of this species make good parents and

rather difficult to distinguish the two sexes. The female's colors are more subdued; her fins are shorter and her body fuller.

At spawning time the male becomes quite impatient and may kill the female if she is not ready. The breeding tank should be large with a number of natural rock formations. The breeders don't tolerate plant life and plants are torn and sand dug up prior to spawning. A corner will be staked out. It is best to allow the male to choose his own mate. The two will grip each other

Even though Firemouths can grow to a length of about six inches, they will spawn at a much smaller size in the aquarium.

well-mated pairs will spawn readily. Prior to spawning, colors deepen, a bright red flush extends over the belly and chin and up into the mouth of the male. When fully mature he will develop a long filament at the tip of his dorsal. Until the fish are mature it is

by the mouth in a tug of war. If this ends in a stalemate, the chances are good the pair will get along well. The spawning site will be meticulously cleaned. Afterwards the male will swim in steadily decreasing circles around his mate. The eggs are laid

on a prepared site and carefully watched over for two to three days until they hatch. The spawnings are large and the young are easily raised. After birth the fry are transported in their parents' mouths to a number of shallow depressions or pits. They are moved around from pit to pit at frequent intervals. If kept at 80°F, they will become free-swimming after three or four days. They are a good size and can eat quite large amounts of freshly hatched brine shrimp and freeze-dried fry food. The parents are best removed at this time.

CONVICT CICHLID
Herichthys nigrofasciatum

The Convict Cichlid is one of the easiest of all cichlids to breed. They make few demands on their environment. They are hearty eaters and will accept any type of commercially prepared food.

A pair of Convicts will do quite well in a 15-gallon tank with numerous flat

Perhaps the most undemanding cichlid to maintain in the aquarium is *Herichthys nigro-fasciatum.*

They are South American natives and prefer aged, slightly acid water with a little sunlight on the tank every day. The water should be about 78 to 85°F. Live foods are preferred, especially tubifex worms, daphnia, and brine shrimp, but they will accept prepared foods as well.

Rams are especially timid if they have a bare tank. They should have some caves and plants for security.

The Ram, *Microgeophagus ramirezi*, requires neutral to acid water in order to function at its best.

rocks and a flowerpot with a wedge cut out. The water should be of moderate hardness with a pH between 6.8 and 8.0, maintaining a temperature between 76 to 80°F. As spawning approaches, their color deepens. They will swim about together in search of a spawning site that they clean thoroughly. After brief courting activities, the eggs are deposited and fertilized, usually inconspicuously. Both parents share in the care of the brood to the delight of the aquarist. These fish are famous for their parental care, choosing to protect their fry even over tempting foods. Breeding pairs will spawn regularly in a large community tank even with other fishes present.

RAM
Microgeophagus ramirezi

Like most of the dwarf cichlids, Rams are beautiful and peaceful.

KRIBENSIS
Pelvicachromis pulcher

Most of the African cichlid species are quite large in size and there are not many that could be considered dwarf species. *Pelvicachromis pulcher* is small enough to come into this category. In addition to their medium size (3 to 4 inches) both sexes are among the most beautifully colored of all freshwater species.

While Kribs are peaceful enough for almost any community aquarium, they do require neutral to slightly acid and soft to medium-hard water. The pH, as usual, is more significant than the hardness. The water temperature should be in the 78 to 82°F range. They accept almost any food, but show a definite preference for tubifex worms and other live foods.

When provided with conditions to their liking, a well-mated pair of these fish will spawn with regularity. One of the most successful methods has been to set a flowerpot with its open end on the bottom. A piece is broken out of the side to permit both fish to swim in and out. The temperature is set at 80°F. They begin exploring in short order and soon you will notice they are spending a great deal of time in the flowerpot. Then one day you will see the male come pell-mell out of the opening, with the indignant female sticking her head out to make sure he doesn't come back. This is the right time to take the male out and give the female a chance to tend her eggs. The eggs hatch in two to three days and the fry become free-swimming three to five days after that.

Pelvica-chromis pulcher, the Krib. The females sport a bright red belly to attract the males.

MBUNA
Pseudotropheus zebra

While *Pseudotropheus zebra* is but one of the many species of mbuna, the mbuna share similar habits and P. zebra is shown as a representative of the mbuna.

Mbuna are cichlids of Lake Malawi, Africa. The term mbuna is simply the chiChewa (the language of Malawi) word for "rock fish." There are presently ten genera of mbuna: *Pseudotropheus, Melanochromis, Petrotilapia, Labidochromis, Cynotilapia, Labeotropheus, Gephyrochromis, Iodotropheus, Genyochromis*, and *Cyathochromis. Aulonocara,* although a member of the non-mbuna haplochromines, by virtue of the rock-inhabiting nature of the species often have been given the status of "honorary" mbuna.

Mbuna in the lake spend a great deal of time grazing for algae and the small crustaceans living therein. They should therefore be provided with both vegetable based and meaty foods. If all is not well with them, however, their colors will fade and they may even stop eating, eventually leading to their demise. These are highly territorial fishes and must be provided with ample rocky territories in the aquarium.

Pseudo-tropheus zebra should be housed only with other hardy fish capable of withstanding its highly aggressive nature. Below is a color variant called Red Top Zebra, from Mbenji Island in Lake Malawi.

Mbuna require many caves and territories within the aquarium. They are quite aggressive and the dominant fish will terrorize the others if they are not given adequate territories. To stifle some of the aggression, mbuna are often stocked at higher densities than usual in the aquarium.

The water in Lake Malawi is quite hard and alkaline with a pH of about 7.8 to 8.2. The addition of special rift lake salts will keep your mbuna happy in the aquarium. The temperature should be about 76 to 80°F.

Your mbuna will happily spawn in the community aquarium, but your fry will have a far better chance at survival if you are able to provide a spawning tank for the parents. The spawning tank should be 20 to 50 gallons. Mbuna exhibit very advanced reproductive behavior in that the female carries the eggs in her mouth— referred to as mouthbrooding—until they hatch, and the fry are mature enough to make it on their own in the world.

The zebras from Lake Malawi come in a multitude of color varieties that are commonly available at your local tropical fish dealer. The zebra above is called the Chilumba Zebra while the one below is called the Orange Zebra.

Every domestically produced strain of Angelfish, *Pterophyllum scalare*, including this black veil variety, originally came from the shorter-finned silver Angelfish.

ANGELFISH
Pterophyllum scalare

The famous Angelfish has been known to aquarium hobbyists for decades. The first specimens to be brought in were considered to be one of the "problem fishes" because it was very difficult to get a pair to mate and spawn. After tank-raised stock became available even novices could breed them.

Mature angelfish are about six inches in length, but have long dorsal and anal fins that can exceed their length. There are many varieties of angelfish: veiltail,

pearlscale, marble, gold, koi, black, blue, etc. It remains only for the hobbyist to choose the type that is most beautiful to his own eye.

Angelfish are quite peaceful for cichlids. They will not usually harass smaller fishes, but if the fish is small enough to eat, beware. They can become a bit nasty with competitors of their own species at breeding time, but other than that, they are docile.

A pair of angelfish will live comfortably in an aquarium of 20 gallons. The best temperature is about 78°F. The water should be neutral to slightly acid and better softer than harder.

DISCUS
Symphysodon spp.

For many years the majestic Discus species were considered impossible to spawn and resisted all efforts made at coaxing them to do so. Many were the trials, and many the failures. Finally some heartening progress was reported. Here and there people told of how their Discus were laying eggs. Mostly the parents would eat the eggs shortly afterward, and so the same technique was resorted to as is standard with Angelfish breeding. The

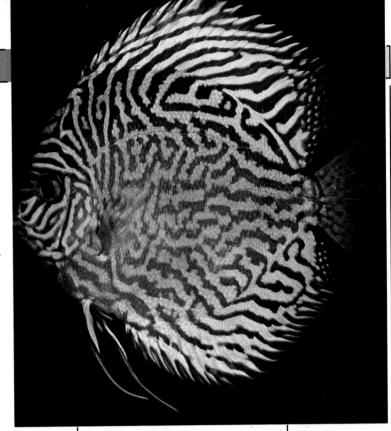

parents were removed after spawning and an airstone was placed so that it would cause a gentle agitation of the water nearby.

This permitted successful hatching and the youngsters grew to a point where their yolk-sacs were absorbed, but here the breeders ran into a brick wall. The fry firmly and steadfastly refused to eat! Finally it was found that Discus babies had to be left in the care of their parents. They would then "graze" off the sides of the breeders and

grow very well on this slime for about a week, after which time they could be fed newly-hatched brine shrimp.

The water for Discus should be soft, about about 2 or 3 DH, but in the absence of soft water, pH is more important, in the range of 6.0 to 6.5, and the temperature should be warm at about 82°F. Discus should be fed all sorts of living and freeze-dried foods to be kept in top condition.

At least 20 percent of their aquarium water should be siphoned off and replaced each week, as Discus cannot tolerate poor water conditions.

Considered by many to be the king of cichlids, Discus, *Symphysodon spp.*, require a little more T.L.C. than most other tropical fish in order to be kept successfully.

A natural way to remove unwanted hydra from an aquarium can be to place two to three Blue Gouramis, *Trichogaster trichopterus*, into such an infested aquarium. In a very short time, the Blue Gouramis will have devoured every hydra in sight.

LABYRINTHFISHES

Members of the labyrinthfish group—Gouramis, Bettas, and Paradisefish—are highly prized by beginners and advanced hobbyists alike. They possess a labyrinth organ that permits them to breathe atmospheric air. Most are bubblenest builders, but some are mouthbrooders. The male builds a nest at the surface of the water and it is here the newly released eggs are incubated. The male guards the eggs and cares for the fry. The males are particularly fierce when guarding their bubblenest and the female should be removed after she has laid her eggs. Her job is done. The male may damage her if she is allowed to remain in the aquarium.

BLUE GOURAMI OR THREE-SPOT GOURAMI
Trichogaster trichopterus

Blue Gouramis from the tropical Far East are robust, prolific, and more hardy and aggressive than most gouramis. They reach an average length of five inches and will begin breeding at three and one-half inches. They should only be kept with other fishes of approximately the same size. Water requirements are flexible; tcmperatures between 70 and 88°F arc quite suitable. They have hearty appetites and accept any freeze-dried or live aquarium food. They are useful for ridding tanks of the dreaded *Hydra* which they find a delicacy.

Their body is silver-blue with a series of incomplete blue bars on the posterior portion of the body. There are two black spots, one on the center of the body, the other on the caudal peduncle. In the common name, "three-spot" includes the eye, which lies on nearly the same line as the body spots. There are several pale pink spots on the fins. The sexes are

similar in appearance, except the males have larger dorsal and anal fins. During the breeding season the males develop a deep midnight-blue color. The females acquire a darker color also. They make a very handsome pair.

DWARF GOURAMI
Colisa lalia

The popular Dwarf Gouramis are native to the vegetation-clogged rivers and streams of India, Bengal, and Assam. They are perfect aquarium fish since they are hardy, peaceful, and quite undemanding. They have an elegant beauty, but unfortunately tend to be shy when placed with aggressive species. Feeding presents little trouble since they will eagerly accept prepared foods which should be supplemented with occasional live foods. *C. lalia* are the smallest of their genus. The males

These dwarf gouramis, *Colisa lalia*, will feel more at home in a thickly planted aquarium.

never exceed two and one-half inches and the females two inches. They are beautiful. The body of the male, which is oval and compressed laterally, is adorned with alternating series of shiny blue-green and red stripes which are also present, although irregular, in the unpaired fins. The dorsal is red with blue dots, while the anal and caudal fins are orange with blue dots and dashes. The elongated extensions of the second rays of the ventrals are an orange-red and are generously provided with taste buds. The throat and belly are indigo. The female is paler; her body is a grayish silver with delicate blue stripes. Her fins are translucent and the dorsal, anal and caudal fins are bordered with red and sprinkled with red flecks.

Her throat and belly are bluish white.

A suitable artificial environment for these fish would include a small tank that is well stocked with floating aquatic plants and receives a generous amount of sunlight. A dark bottom is recommended, since it shows their lovely colors off to best advantage. Normal tap water, which varies from neutral to slightly acid, may be used. Aeration is not essential because of their labyrinths, but the temperature should be relatively warm (75 to 81°F).

HONEY GOURAMI
Colisa chuna
This very attractive little fish is a fairly recent introduction to the aquarium hobby. It comes from India. The Honey Gourami, as the aquarium

hobbyists were quick to call it, should never be judged by the pale colors one sees in a dealer's bare tank. It is rather shy by nature, and a bare tank is a frightening experience to a shy fish. But put them in a well-planted tank and the magic soon begins. The female's yellow color deepens, and there is a definite dark stripe from the gills to the caudal base. The male gets the real colors. His body is a shade deeper and at spawning time his face, throat, and belly become very dark blue. The dorsal fin is slightly orange.

Water conditions are not at all critical as long as they are not extreme and cleanliness prevails. They like warm water, 78 to 82°F. A number of plants may be planted in the gravel bottom and will give them the security they need. They are very fond of live or frozen *Daphnia* and brine shrimp, fresh and freeze-dried tubifex worms, and white worms in addition to prepared food as a staple.

Above: The male Honey Gourami tends the bubblenest. **Below:** The male Honey Gourami wraps his body around the female's during spawning.

The Paradise Fish, *Macropodus opercularis*, is a highly adaptable fish and makes a wonderful aquarium resident.

PARADISE FISH
Macropodus opercularis

Paradise Fish are native to the stagnant, muddy streams and rice paddies of Southern China and Formosa. There are numerous opinions as to their disposition. Although they can occasionally live in harmony with other fishes, they are perpetually hostile toward their own kind and have an undesirable habit of making sudden lunges at members of their own species, butting rather than nipping them. They are able to adjust to extremes in temperatures from 50 to 90°F and can be kept in an unheated aquarium or an outdoor pool during the warmer months. High temperatures tend to shorten their life. They seem to do best at around 65 to 70°F while 70 to 75°F is sufficient for breeding. Water conditions are not critical, and neutral values are quite acceptable.

They have enormous appetites and should be fed copious amounts of live, frozen, freeze-dried, and prepared foods. They are full grown at about three inches but will start breeding at two and one-half inches.

They are quite handsomely colored. A brownish green background is transversed with alternating dark blue and red metallic bands. The fins are blue with light stripes. The tail, which is red with green bands and dotted with blue spots, is forked into two long blue filaments. The dorsal is large and blue with white margin and spots. The anal fin is blue anteriorly becoming red posteriorly. The sexes are difficult to distinguish before maturity. The male is the more aggressive with deeper coloration and larger

fins. He is especially gorgeous at breeding time when he sports red, peacock blue, orange and emerald green in tasteful combinations. The female is duller having only the red transverse bands. There are three color variations of this species: red, black and albino.

PEARL GOURAMI
Trichogaster leeri

Pearl Gouramis are one of the most elegant and dignified of all tropical aquarium fishes. They are timid and gentle, and ideal for a community tank. These fish may reach a length of five inches but average about four inches. They are undemanding and will politely accept most aquarium foods, although their diet should include some live or freeze-dried foods. They are exceptionally fond of worms and thrive on them. They are very useful for ridding the tank of unwanted *Hydra*. Remember that their mouths are relatively small and upturned and is best suited for collecting surface foods.

They are native to the waters of Thailand (the Malayan Peninsula), Sumatra, and Borneo. Provide them with a large well-planted tank of at least 20 gallons. Water requirements are flexible but neutral water of medium hardness and a average temperature of around 78°F is quite suitable. They seem to prefer warm water.

They are beautiful, graceful fish. Their coloration is variable. The body usually has delicate

The Pearl Gourami, *Trichogaster leeri*, is perhaps the most strikingly attractive gourami that is readily available to hobbyists.

The majestic Siamese Fighting Fish, *Betta splendens*, should not be kept with other fish that have an inclination to nip their long, flowing fins.

silver-blue luster. There is a fine mosaic of pearly gray flecks and a dark irregular horizontal line stretching from the mouth to the base of the tail. The sexes are easily distinguished. The male has a golden orange throat and belly which become almost a glowing red during the breeding season. His dorsal is long, his body is strongly compressed laterally and his long thread-like ventrals are a deep red. The female is heavier and her colors more subtle. Her fins are shorter and rounder.

SIAMESE FIGHTING FISH
Betta splendens

The Betta, or Siamese Fighting Fish, is among the most famous of all aquarium fishes. From the original wild specimens that came from southeastern Asia, aquarists have bred many shades of long-finned, three-inch beauties that could vie successfully in color and grace with many of our butterflies. The Thai natives, on the other hand, are a little more practical. Instead of breeding for lovely colors, they try to produce males that will fight bitterly when placed with another of their own sex. Such combats attract spectators who wager money on the outcome, and it is readily understandable how the owner of a good battler could have himself a fairly valuable bit of property for a time.

Bettas are happy with very little aquarium upkeep. Keep the water temperature close to 80°F and clean and your Betta will be fine. A few sprigs of plants floating at the surface are an asset. They don't even require filtration. A small tank is all that is necessary if you are going to keep a male and a female or two alone together. The males build bubblenests even when they are bachelors so the presence of females will encourage their efforts even more. Feed your Bettas a variety of live and prepared foods and they will thrill you with their grace and beauty.

Bettas are ideal in the community tank as long as the other fish don't harass them. Beware of keeping them with nippy species that find their long, flowing fins irresistible.

TETRAS

Most of the tetras are South American natives. The tetras are all egglayers. Mostly the eggs are scattered or sprayed wherever a handy bunch of plants is found. The tetras are tremendously fertile if the breeding stock is properly conditioned and fully grown. In fact, the danger actually lies in not utilizing

This variety of *Betta splendens* is called a marbled doubletail.

your egg-filled females often enough. Mature females living without mature males can become egg-bound and subsequent breeding is difficult if not impossible.

The family Characidae contains a tremendous number and variety of tetras. From Cardinal Tetras to Piranhas, the Characins are fascinating and desirable aquarium fishes.

If kept in a school and maintained in soft water, these Cardinal Tetras, *Paracheirodon axelrodi*, will display optimal health and coloration.

CARDINAL TETRA
Paracheirodon axelrodi

The Cardinal Tetra, native of the Upper Rio Negro and its tributaries, is probably the most lovely of all tropical aquarium fishes. They are peaceful and active and are best kept together in a group of at least six. They are hardy and not extremely susceptible to disease. Feeding is no problem, they will eat all types of freeze-dried food, live, flake, and frozen food.

Unfortunately it is not always easy to provide the soft, acid water Cardinals require for spawning, so the majority of the Cardinals in our aquaria are actually wild stock imported for the aquarium trade.

Cardinal Tetras grow to about 2 inches and are quite similar to the Neon Tetra in form and color but they are slimmer and have a brighter red. Their backs are brown and the upper part of the body has a brilliant iridescent bluish green stripe. The lower part of the body is a flaming red and the belly is white.

Their native waters are brown, rich in humic acid and low in mineral content. For this reason they do best in soft (2 to 3 DH), well aged, acid (pH 5 to 6.8) water that is very clear. They benefit tremendously if their water is filtered over peat for its humins, tannins, and acidifying property.

FLAME TETRA
Hyphessobrycon flammeus

The Flame Tetras, from Brazil in the vicinity of Rio de Janeiro, are very prolific and easily bred. They average about one and one-half inches in length and because of their size should only be kept with other small fishes, preferably in schools of at least a half dozen. An aquarium of 10 gallons or larger that is well stocked with bushy plants is recommended. They are not timid and are on constant display racing up and down the tank. They are not demanding about the water conditions and do very well in water that is soft and neutral to slightly acid with a temperature of 76 to 80° F. They accept most commercially prepared foods but need occasional live foods to stay in good condition.

They are easily brought into breeding condition with generous feedings of live, frozen, and freeze-dried foods. A trio, two males to each female, is ideal. The males will chase the female about madly. She finally darts in among the plant bundles where she quickly deposits about 10 eggs. A male will then come up alongside her to fertilize them. The female shoots out and again begins to swim actively to

The Flame Tetra, *Hyphesso-brycon flammeus*, will do quite well in 10- to 15-gallon aquariums.

The gorgeous Rummy-Nose Tetra, *Hemigrammus rhodostomus*, like most other small tetras, needs to be kept with others of its own kind in order to feel secure in an aquarium.

and fro until she again is escorted back into the thickets. Courting is continued until 100 to 200 semi-adhesive eggs have been laid. These are small and almost clear. The parents should be removed immediately after spawning or they will gobble up all the eggs they can find. The eggs will hatch in two to three days depending upon the temperature. Algae is a good first food as well as very fine powdered food or infusoria. They reach maturity in about five months.

RUMMY-NOSE TETRA
Hemigrammus rhodostomus

The Rummy-nose Tetra is a native of the lower Amazon region and is peaceful and at home with most fishes of its own size. When kept in an aquarium that is well established and well planted, a small school of these fish flit around always active and alert.

Rummy-noses are not the easiest of the tetras to breed and once the fry arrive, they are not the easiest to raise. However, if provided with aged, soft, and very acid water in the 80°F range, the results can be gratifying.

A diminutive fish, the Rummy-nose measures about two inches. The glowing red nose is a sure sign that all is well with this species. At breeding time, the females fill with roe and the red nose of the males becomes a beacon.

Fish Health

We all want our fishes to stay healthy for a long time, but a fish like any other living creature is subject to illness, injury, and, sharing the eventual fate of all living things, will someday perish. However, every trauma need not be fatal to the hardy fishes you are keeping in your aquarium. Even though these little animals sometimes go belly-up for no apparent reason, many of the common ailments suffered by fishes are easily avoided. Let us discuss some of the problems you could encounter with the health of your fishes and propose solutions.

While it is laudable to know all the cures for the various illnesses, there is one nostrum that is effective for most maladies—prevention. How does one help prevent fish disease? By being a good aquarist.

STRESS

Just about every fish undergoes considerable stress before it is safely ensconced in your home aquarium. Stress is a major factor of disease and death in your fishes. The most obvious stressful situations are those that cause fright, discomfort, or pain. Handling and capture, while sometimes necessary, should be avoided, since they are primary sources of stress.

If you can keep up with the responsibilities of aquarium maintenance and closely observe your fishes on a regular basis, you can expect your aquarium set up to function properly as seen here.

Territorial behavior is another source of stress in fishes. It has been determined that the aggressive behavior of dominant fishes can cause severely frayed fins and darkened bodies in the chronically stressed victims. If you notice one fish hanging out in a corner constantly and being harassed by another whenever it moves, you know this fish is under stress. Either rearrange the rockwork or move the fish to another tank, otherwise the fish will be dead in a matter of days or hours.

Confinement, anesthesia, bad water quality, transport, and changes in water temperature are all stress-producing factors in fishes. These are things to be avoided whenever possible.

While no fish can reach our tanks without undergoing some stress, remember that this is not good for your fishes and will certainly cause problems (just like in humans) if permitted to continue. Studies have shown that proper conditioning is one valuable way to reduce the effects of stress. Another valuable method is to ensure that care is taken to minimize the number of stress inducers at one time, such as handling combined with a change in temperature, pH, or poor water quality.

Whether our new fishes are wild or cultivated, they have been captured, stored, crowded, chilled, heated, and transported several times. They are usually underfed (if fed at all) and

One must be careful not to allow excessive fights to go on, since they will inevitably lead to the death of the weaker fish.

only achieve temporary comfort in the dealer's tank. So when we get our prizes home, it is necessary to bear in mind that these poor creatures have been through quite a lot and need to be treated with consideration. When you first bring them home and before you release them, slowly mix the aquarium water with the bag water (in a clear container) in order to equalize the chemistry.

Your first feeding attempts may be frustrating, not necessarily because the fishes are sick, they are just disoriented, and even though hungry they may not be too interested in food.

POISONING

Poisoning is a common cause of death in fishes, even before disease processes have had time to manifest themselves. Chlorine and chloramine

Newly acquired fish need to be given some time to recuperate and adjust to their new surroundings. The White Clouds, *Tanichthys albonubes*, shown here are exploring a new-to-them tank.

Products for treating the aquarium's water to make it safer for the fish to live in are available at most pet shops. Photo courtesy of Hagen.

Oxygen deficiency can easily be remedied by putting to use an air pump with an airstone attached.

must be removed from the water. It is simple enough to age the water to remove the chlorine and treat with a neutralizer (sodium thiosulphate) for chloramine. Another poison that is sometimes overlooked is actually produced by the fishes themselves. The ammonia and nitrite building up daily in your tank can be just as poisonous as DDT and the only treatment for this is prevention. Perform the routine water changes and you will not have to worry about your crystal clear but biologically filthy water weakening your fishes and making them targets for any passing pathogen.

OXYGEN DEFICIENCY

Oxygen deficiency is also contributory to disease processes in our fishes. Regardless of what type of fishes we keep, oxygen is required for respiration. Warmer water has a lower oxygen-carrying ability than cooler water.

The causes of the diseases to which fishes are prone are many, but in general they are caused by bacteria, parasites, viruses, and fungi. It is almost impossible for an aquarist to make an abso- lutely ac- curate

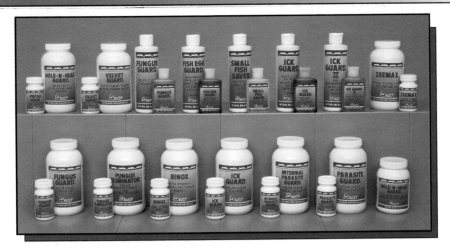

diagnosis without considerable knowledge, experience, and equipment, but there are some conditions that can be assumed from symptoms.

DISEASES

Ich

Ichthyophthiriasis, commonly called "ich," is a disease caused by a protozoan that burrows into the skin of the fish and causes a little white blister. At first the fish rubs its body on rocks, plants, and other objects in the tank, as if scratching itself. Then you can see small white spots on the fins and later on the body of the fish.

Ich is very contagious and if present in one tank can be transferred to another by using the same net, etc. for both. So, as with most diseases, hygiene is a good preventative measure. Wash your hands after working with one tank and before going to another. Use separate nets for each tank. Never trade plants from a diseased tank to a healthy tank. The same goes for fishes. If a fish looks healthy, but has been in a tank with problems, don't put it into a healthy tank. Chances are it is carrying trouble, so cure the whole tank before transferring anything from it to another.

The therapy for ich is to elevate the temperature of the water to about 85 or 86°F. This speeds up the life cycle of the parasite and kills the young. Malachite green, available in many packaged ich cures, is the treatment of choice. Use exactly according to package directions.

Many different remedies, preventives and tonics are available at pet shops. Photo courtesy of Jungle Laboratories.

Velvet is another disease often seen on aquarium fish; however, it seems to be more common in soft water than in hard water.

Shimmies

Shimmies is a condition rather than a disease. The fish swims but doesn't get anywhere, wagging its body from side to side. Ich, indigestion, chills, etc., are possible causes of this condition. A water change and elevation of the temperature usually corrected this malady.

Velvet

Velvet *(Oodinium)* is a parasite that attacks the gills and skin of affected fish. It is recognized by a yellow-brown film that starts near the dorsal fin and spreads to the rest of the body in a velvet-like covering. Acriflavine, methylene blue, or formalin are used in treating this disease.

Salt Baths

A progressive salt bath is performed in a hospital tank that contains no plants or decorations. Salt is added in the proportion of two level teaspoons per gallon of water on the first day. Dissolve the salt in water before adding it to the fish tank. The second day add two more teaspoons per gallon, and one teaspoon per gallon on the third day. Marine salt is preferred over table salt. The progressive salt bath is used for several illnesses including Saprolegniasis and pop-eye.

HOSPITAL TANK

The hospital tank is a necessary accessory when you have an outbreak of any kind of illness in your community tank. Adding medications to a well-established tank because one or even several fishes exhibit signs of disease is not a well-advised thing to do. You never want to dose healthy fishes. Some of the medications used in the aquarium require that some or all of the water be totally changed, which could be difficult in the community tank. Also, the disease could become well-established in the tank in the time it takes to attempt a cure.

Suggested Reading

For your further information, consult any of these highly informative T.F.H. tropical fish books.

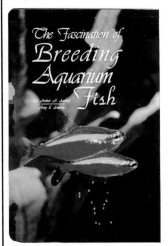

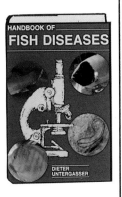

TS-185, 448 pages with over 500 full-color photos

H-1077, 1024 pages with over 7,500 full-color photos

TS-123, 160 pages with over 100 full-color photos

Index

Ammonia, 23–24, 50
Angelfish, 76
Apisto, 67–68
Apistogramma agassizi, 67–68
Barbs, 62–63
Barbus tetrazona, 63
Betta splendens, 84–85
Blooming Algae, 51
Blue Gourami, 78–79
Botia macracantha, 49
Box Filters, 10–11
Brachydanio rerio, 62
Bronze Catfish, 64–65
Canister Filter, 13
Cardinal Tetra, 86
Catfishes, 64–66
Chloramine, 24, 49–50
Chlorine, 24, 49–50
Cichlids, 67–77
Colisa chuna, 80–81
Colisa lalia, 79–80
Convict Cichlid, 71–72
Corydoras aeneus, 64–65
Corydoras imitator, 39

Corydoras paleatus, 65–66
Corydoras reticulatus, 33
Cycling, 30–32
Diatomaceous Earth Filter, 51
Discus, 76–77
Diseases, 93–94
Dwarf Gourami, 79–80
Filters, 9–10
Firemouth Cichlid, 69–71
Flame Tetra, 87–88
Fossorochromis rostratus, 37
Gravel Vacuum, 51–52
Gravel, 18
Guppies, 53–55, 58–60
Heaters, 7–9
Hemichromis lafalili, 68–69
Hemigrammus rhodostomus, 88
Herichthys meeki, 69–71
Herichthys nigrofasciatum, 71–72
Honey Gourami, 80–81
Hospital Tank, 95
Hyphessobrycon flammeus, 87–88

Megal-amphodus roseus, one of the "phantom" tetras. Photo by Wolfgang Sommer.

Hypostomus scaphyceps, 66
Hypostomus sp., 66
Jewel Cichlid, 68–69
Kribensis, 72–73
Labyrinthfishes, 78–85
Lighting, 15–18
Macropodus opercularis, 44, 82–83
Mbuna, 74–75
Microgeophagus ramirezi, 72
Mollies, 153–57, 59–61
Nitrates, 23–24, 50
Nitrites, 23–24, 50
Outside Power Filters, 11
Paracheirodon axelrodi, 86
Paracheirodon innesi, 45
Paradise Fish, 82–83
Pearl Gourami, 83–84
Pelvicachromis pulcher, 72–73
Peppered Corydoras, 65–66
Platies, 55, 58–60, 53Pleco, 66
Poecilia latipinna, 55–56
Poecilia reticulata, 54–55
Poecilia sphenops, 1, 57
Poecilia velifera, 55–56, 59
Pseudotropheus lombardoi, 35
Pseudotropheus zebra, 74–75
Pterophyllum scalare, 34, 76
Pumps, 15
Ram, 72
Rummy-Nose Tetra, 88
Salt Baths, 94–95
Siamese Fighting Fish, 84–85
Sponge Filter, 12–13
Swordtail, 57–58, 59–60
Symphysodon spp., 49, 76–77
Tanichthys albonubes, 62–63
Tetras, 85–88
Thermometers, 9
Tiger Barb, 63
Trichogaster leeri, 83–84
Trichogaster trichopterus, 47, 78–79
Undergravel filter, 12
Water Chemistry, 21–23
White Cloud, 62–63
Xiphophorus helleri, 57–58
Xiphophorus maculatus, 43, 53
Xiphophorus variatus, 55
Zebra Danio, 62